T0315126

**Using Predictive Analytics to Improve
Healthcare Outcomes**

Using Predictive Analytics to Improve Healthcare Outcomes

Edited by

John W. Nelson
Healthcare Environment

Jayne Felgen
Creative Health Care Management

Mary Ann Hozak
St. Joseph's Health

Registered Office
John Wiley & Sons, Inc., 111 River Street, Hoboken, NJ 07030, USA

Editorial Office
111 River Street, Hoboken, NJ 07030, USA

For details of our global editorial offices, customer services, and more information about Wiley products visit us at www.wiley.com.

Wiley also publishes its books in a variety of electronic formats and by print-on-demand. Some content that appears in standard print versions of this book may not be available in other formats.

Library of Congress Cataloging-in-Publication data applied for

ISBN 978-1-119-74775-8 (hardback)

Cover Image: © Kaikoro/Adobe Stock (adapted by Healthcare Environment)
Cover Design by Wiley

This book is dedicated to the memory of John Lancaster, MBE, who died on June 20, 2020. A dedicated nurse, John felt it vital to capture the importance of caring in a scientific way. He was delighted to have been involved in the research described in this book. John will be fondly remembered by family, friends, colleagues, and patients for his humor, kindness, and compassion, as well as the admirable way he lived with cancer, facing death with a dignity sustained by the Catholic faith that was so important to him.

Contents

Contributors

Kate Aberger
Medical Director
Division of Palliative Care and
Geriatric Medicine
St. Joseph's Health
Paterson, NJ, US

Pauline Anderson-Johnson
Lecturer
University West Indies School
of Nursing
Mona, Jamaica

Alba Barros
Professor
Federal University São Paulo – Escola
Paulista de Enfermagem
São Paulo, Brazil

Jacqueline Brown
Clinical Educator
Golden Jubilee National Hospital
Clydebank, Scotland

Dawna Cato
Chief Executive Officer
Arizona Nurses Association
Mesa, AZ, US

Sally Dampier
Professor
Confederation College
Thunder Bay, Ontario, Canada

Inge DiPasquale
Manager
Division of Palliative Care and
Geriatric Medicine
St. Joseph's Health
Paterson, NJ, US

Melissa D'Mello
Congestive Heart Failure Coordinator
St. Joseph's Health
Paterson, NJ, US

Ana Esteban
Associate Director Quality Regulatory
Compliance
Columbia Doctors – The Faculty
Practice Organization of Columbia
University Irving Medical Center
New York, NY, US

Jayne Felgen
President Emeritus and Consultant
Creative Health Care Management
Minneapolis, MN, US

Irit Gantz
Coordinator
Woman-Health Division School
of Nursing
Meir Hospital
Kfar-Saba, Israel

Kary Gillenwaters
Chief Executive Officer
Solidago Ventures and Consulting
Elk River, MN, US

Sebahat Gözüm
Dean, School of Nursing
Professor, Department of Public
Health Nursing
Akdeniz University
Antalya, Turkey

Lidia Guandalini
Cardiology Nurse
Federal University São Paulo – Escola
Paulista de Enfermagem
São Paulo, Brazil

Alicia House
Executive Director
Steve Rummler Hope Network
Minneapolis, MN, US

Mary Ann Hozak
Administrative Director
Department of Cardiology
St. Joseph's Health
Paterson, NJ, US

Michal Itzhaki
Senior Lecturer
Department of Nursing
Tel Aviv University
Tel Aviv, Israel

Benson Kahiu
Nurse Manager
Mount Sinai Health
East Orange, NJ, US

Ayla Kaya
Research Assistant Director
Pediatric Nursing
Akdeniz University
Antalya, Turkey

Gay L. Landstrom
Senior Vice President and Chief
Nursing Officer
Trinity Health
Livonia, MI, US

Marissa Manhart
Performance, Safety, and Improvement
Coordinator
St. Joseph's Health
Paterson, NJ, US

John W. Nelson
Chief Executive Officer
Healthcare Environment
St. Paul, MN, US

Tara Nichols
Chief Executive Officer and Clinician
Maters of Comfort
Mason City, IA, US

Kenneth Oja
Research Scientist
Denver Health

Assistant Professor
University of Colorado
Denver, Colorado, US

Dawna Maria Perry
Chief Nursing Officer
Thunder Bay Regional Health
Science Center
Thunder Bay, Ontario, Canada

Lance Podsiad
Manager
Helios Epic
Nurse Manager
Henry Ford Health System
Detroit, Michigan, US

Karen Poole
Associate Professor
Lakehead University School
of Nursing
Thunder Bay, Ontario, Canada

Rebecca Smith
Writer/Editor
Minneapolis, MN, US

Susan Smith
Chief Executive Officer
Choice Dynamic International
Leeds, England

Kay Takes
President
Eastern Iowa Region of MercyOne
Dubuque, IA, US

Patricia Thomas
Manager – Associate Dean
Nursing Faculty Affairs
Wayne State University
College of Nursing
Detroit, MI, US

Anna Trtchounian
Emergency Medicine Resident
Good Samaritan Hospital
Medical Center
West Islip, Long Island, NY, US

Sebin Vadasserril
Manager
Innovative Nursing Practice
and Quality
St. Joseph's Health
Paterson, NJ, US

Linda Valentino
Vice President
Nursing Operations
Mount Sinai Hospital
New York, NY, US

Dominika Vrbnjak
Assistant Professor
University of Maribor Faculty of
Health Sciences
Maribor, Slovenia

Josephine (Jo) Sclafani Wahl
Associate Director
BRG/Prism
MI, US

Jacklyn Whitaker
Nurse Manager
St. Joseph's Health
Paterson, NJ, US

Theresa Williamson
Associate Nurse Director
Golden Jubilee National Hospital
Clydebank, Scotland

Foreword

John W. Nelson and his colleagues are to be congratulated for creating this distinctive book. A very special feature of the book is the use of predictive analytics to explain, amplify, and validate caring theory. All too often, publications focusing on methods such as predictive analytics ignore the theoretical frameworks that guide the collection of data to which analytics are applied. The reader is then left with the thought, "Perhaps interesting results, but so what?" This book provides the answer to "so what?" by presenting the very interesting results, within the contexts of caring theory, specifically Relation-Based Care®, the Caring Behaviors Assurance System©, and Watson's Theory of Transpersonal Caring.

The book's content emphasizes quality improvement, which might be considered the most appropriate application of predictive analytics in healthcare. Determining how, when, and why to improve the quality of healthcare, as a way to improve individual-level and organization-level outcomes, is a major challenge for all healthcare team members and researchers. Theory-based predictive analytics is an innovative approach to meeting this challenge.

A challenge for the authors of the chapters of this book, and for its readers, is to determine the most appropriate place for theory in the triad of data, theory, and operations. Given my passion for the primacy of theory, I recommend that the starting point be theory, which determines what data is to be collected and how the data can be applied to operations.

The case studies that make up the several chapters of Sections Two and Four of this book, the contents of which are as interesting as they are informative, help readers to appreciate the value of theory-based predictive analytics. The case studies, which range from individual-level problems to department-level problems to health system-level problems, underscore the wide reach of theory-based predictive analytics.

I contend that the ultimate challenge of predictive analytics will be to carry out the theoretical and empirical work needed to test the book editors' claim, in the Preface of this book, that the same formulas helping people in the trucking and mining industries to create profiles of risk that enable them to prevent unwanted outcomes before they happen, can be applied successfully to improve healthcare outcomes. Meeting this challenge will undoubtedly extend the knowledge of our discipline, which many of us now refer to as nursology (see https://nursology.net).

Jacqueline Fawcett, RN, PhD, ScD (hon), FAAN, ANEF
Professor, Department of Nursing, University of Massachusetts Boston
Management Team Facilitator, https://nursology.net

Preface: Bringing the Science of Winning to Healthcare

John W. Nelson

A few years before the publication of this book, I attended an international mathematics conference for research in simulation studies and predictive analytics. Out of more than 300 attendees, there was only one other attendee from healthcare. For three days there were presentations by researchers from the fields of logistics (trucking) and mining, reviewing how they used predictive analytics and simulation to proactively manage outcomes related to productivity and company output. Surely, I thought, the same kinds of mathematical formulas presented by the truckers and miners could be used in healthcare to move us from reactive use of data to a proactive approach.

Currently, hospitals evaluate outcomes related to falls and infections using hindsight-based analytics such as case studies, root cause analyses, and regression analyses, using retrospective data to understand why these outcomes occurred. Once the underlying causes for the outcomes are identified, the organization creates action plans for improving the outcomes. The problem with this process is that retrospective data provides only hindsight, which does nothing to create a profile of current or future risk. Healthcare organizations typically stop short of supporting *pro*spective management of the data, which would allow for the collection of meaningful data about real-life trends and what is actually happening in practice right now. Conversely, the truckers and miners at the conference showed how predictive analytics can be used to study risk for the purpose of managing unwanted outcomes *before* they occur. Since I am both a data scientist and a nurse, I could see clearly that the formulas from the math conference could apply to healthcare; all you would have to do is specify the models.

This book is about how analytics—mostly predictive analytics—can be used to improve outcomes in healthcare. This book also reveals how good data, derived from good theory, good measurement instruments, and good data collection processes has provided actionable information about the patient, the caregiver, and the operations of care, which have in turn inspired structure and process changes that saved millions of dollars while improving the experience of both patients and providers.

Organizations that have embraced predictive analytics as a central part of operational refinement include Amazon, IBM (Bates, Suchi, Ohno-Machado, Shah, & Escobar 2014), Harrah's casino, Capital One, and the Boston Red Sox (Davenport 2006). In his 2004 book (and the 2011 film), *Moneyball*, Michael Lewis, documents an example of how in 2002 the Oakland A's professional baseball team, which had the lowest payroll in baseball, somehow managed to win the most games. This paradox of winning the most games despite having the skimpiest budget in the league was due to an assistant general manager who used a baseball-specific version of predictive analytics called sabermetrics to examine what combination of possible recruits would reach first base most reliably, and would therefore result in the team winning the most games. These recruits were not the most obvious players—in fact, they were not considered by almost anyone to be the *best* players. It was only predictive analytics that made them visible as the *right* players to comprise this winning team.

If predictive analytics can help a team win more games, why couldn't they help patients heal faster? Why couldn't they help clinicians take better care of themselves? Why couldn't predictive analytics be used to improve every outcome in healthcare?

As a data scientist and operations analyst, it is my job to present data to healthcare leaders and staff members in a way that allows them to easily understand the data. Therefore, it is the job of this book to help people in healthcare understand how to use data in the most meaningful, relevant ways possible, in order to identify the smartest possible operational improvements.

For decades, the three editors of this book have been conducting research to measure some of the most elusive aspects of caring. This book provides instructions and examples of how to develop models that are specified to the outcomes that matter most to you, thereby setting you up to use predictive analytics to definitively identify the most promising operational changes your unit or department can make, *before* you set out to change practice.

List of Acronyms

A&O	Alert and oriented
ACCF	American College of Cardiology Foundation
ACE	Angiotensin-converting enzyme
ACEI	Angiotensin-converting enzyme inhibitor
AGFI	Adjusted goodness of fit index
AHA	American Heart Association
AMI	Acute myocardial infarction
ANEF	Academy of Nursing Education Fellow
ANOVA	Analysis of variance
APN	Advanced practice nurse
ARB	Angiotensin receptor blockers
ARNI	Angiotensin receptor-neprilysin inhibitor
ASAM	American Society of Addiction Medicine
Auto-Falls RAS	Automated Falls Risk Assessment System
BNP	Brain natriuretic peptide
BSN	Bachelor of science in nursing (degree)
BUN	Blood urea nitrogen
CAC	Coronary artery calcium
CAD	Coronary artery disease
CARICOM	Caribbean Community (a policy-making body)
CAT	Caring Assessment Tool
CBAS	Caring Behaviors Assurance System$^{©}$
CDI	Choice Dynamic International
CFI	Comparative fit index
CCU	Coronary care unit
CDC	Centers for Disease Control
CEO	Chief executive officer
CES	Caring Efficacy Scale
CFS	Caring Factor Survey$^{©}$

CFS-CM	Caring Factor Survey – Caring of Manager
CFS-CS	Caring Factor Survey – Caring for Self
CFS-CPV	Caring Factor Survey – Care Provider Version
CFS-HCAHPS	Caring Factor Survey – hospital consumer assessment of healthcare providers and systems (a 15-item patient/ provider survey)
CKD	Chronic kidney disease
CL	Central line
CLABSI	Central line-associated bloodstream infection
CMS	Centers for Medicare and Medicaid Services
CNA	Certified nursing assistant
CNO	Chief nursing officer
CNS	Clinical nurse specialist
COPD	Chronic obstructive pulmonary disease
CPM	Clinical Practice Model
CPR	Cardiopulmonary resuscitation
CPS	Caring Professional Scale
CRT	Cardiac resynchronization therapy
CRT-D	Cardiac resynchronization therapy defibrillator
CRT-P	Cardiac resynchronization therapy pacemaker
CVA	Cerebrovascular accident
CQI	Continuous quality improvement
CVC	Central venous catheter
CHCM	Creative Health Care Management®
DNP	Doctor of nursing practice
DNR	Do not resuscitate
DNR-B	Allows aggressive care, but not to the point of cardiopulmonary resuscitation
DVT	Deep vein thrombosis
ED	Emergency department
EF	Ejection fraction
EFA	Exploratory factor analysis
EKG	Electrocardiogram
EKG QRS	A segment of the EKG tracing
ELNEC	End-of-Life Nursing Education Consortium
EMR	Electronic medical record
ESC	European Society of Cardiology
FAAN	Fellow American Academy of Nursing
FTE	Full-time employee
GFR	Glomerular filtration rate

GLM	General linear model
GPU	General patient-care unit
GWTG	Get With The Guidelines (measurement tool)
HAI	Hospital-acquired infection
HCA	Healing Compassion Assessment
HCAHPS	Hospital Consumer Assessment of Healthcare Providers and Systems
HEE	Health Education of England
HES	Healthcare Environment Survey (measurement instrument)
HF	Heart failure
HMO	Health maintenance organization
ICD	Implantable cardioverter defibrillator
ICU	Intensive care unit
IRB	Institutional review board
IV	Intravenous or information value
I_2E_2	Inspiration, infrastructure, education and evidence
IOM	Institute of Medicine
KMO	Kaiser–Myer–Olkin (mathematical tool)
LOS	Length of stay
LCSW	Licensed clinical social worker
LVN	Licensed vocational nurse
LVSD	Left ventricle systolic dysfunction
MAT	Medication-assisted treatment
MBE	Member of the British Empire
MICU	Medical intensive care unit
MFS	Morse Falls Scale
ML	Machine learning
MSN	Master of science in nursing (degree)
MRN	Medical record number
NA	Nursing assistant
NHS	National Health Service
NHSN	National Healthcare Safety Network
NICE	National Institute of Health and Care Excellence
NNMC	Nichols–Nelson Model of Comfort
NT-proBNP	N-terminal pro-brain natriuretic peptide
O_2	Oxygen
OT	Occupational therapist/occupational therapy
OUD	Opioid use disorder
PC	Palliative care
PCA	Patient care attendant

PCI	Percutaneous coronary intervention
PICC	Peripherally inserted central catheter
PMT	Pacemaker mediated tachycardia
PN	Pneumonia
PCLOSE	An indicator of model fit to show the model is close-fitting and has some specification error, but not very much.
POLST	Physician orders for life sustaining treatments
PPCI	Professional Patient Care Index
PR	Pregnancy related
PSI	Performance and safety improvement
PSI RN	Performance and safety improvement registered nurse
QI	Quality improvement
QRS	(See EKG QRS)
R	A programming language for statistical computing supported by the R Foundation for Statistical Computing.
R4N	Name of medical unit
R6S	Name of medical unit
RAA or R+A+A	Responsibility, authority, and accountability
RBBB	Right bundle branch block
RBC	Relationship-Based Care®
RMC	Recovery management checkups
RMSEA	Root mean square error of approximation
RN	Registered nurse
SAMSA	Substance Abuse and Mental Health Services Administration
SAS	Statistical Analysis System is a software system for data analysis
SBP	Systolic blood pressure
ScD	Doctor of science
SCIP	Surgical care improvement project
SCN	Senior charge nurse
SCU	Step-down unit
SEM	Structural equation model
SPSS	Statistical Package for the Social Sciences is a software system owned by IBM (International Business Machines)
SRMR	Standardized root mean square residual
STS	Sociotechnical systems (theory)
ST-T	Segment of the heart tracing in an electrocardiograph
SUD	Substance use disorder
TIA	Transient ischemic attack
TIP	Treatment improvement protocols

TLC	Triple lumen catheter
TTE	Transthoracic echocrdiogram
UTD	Unable to determine
VS	Vital sign
VS: SBP	Vital sign: systolic blood pressure
VS: DBP	Vital sign: diastolic blood pressure

Acknowledgments

First, the three editors of this book would like to acknowledge our developmental editor, Rebecca Smith, who has made the inaccessible, complex concepts of data analytics simple to understand and exciting to contemplate.

Secondly, we would like to acknowledge all the analysts and mathematicians from other disciplines who have enthusiastically and humbly shared their knowledge of mathematics and how it is applied in science. We have been inspired by the depth and breadth of what you know and by your eagerness to learn from others. The lead editor would also like to ask the indulgence of all of the mathematicians, analysts, and scientists who will read this book, as you encounter moments in this book where brevity and simplicity have taken precedence over thorough scientific explanations. In an effort to make this book accessible to a lay audience, much of the technical talk has been truncated or eliminated.

Thirdly, we acknowledge the visionary leaders who had the courage to step out and measure what matters—behavior and context. Without your understanding that data beyond frequencies was needed, the ability to use predictive analytics to improve healthcare outcomes would still be an elusive dream.

Finally, the editors of this book acknowledge all the staff members who took part in these studies. Every one of you made each model of measurement better, and you played a vital part in producing the groundbreaking findings in this book. Without you, this book would not exist.

Section One

Data, Theory, Operations, and Leadership

1

Using Predictive Analytics to Move from Reactive to Proactive Management of Outcomes

John W. Nelson

For predictive analytics to be useful in your quest to improve healthcare outcomes, models for measurement must reflect the exact context in which you seek to make improvements. Data must resonate with the staff members closest to the work, so that action plans premised on the data are specific, engaging, and instantly seen as relevant. This chapter provides 16 steps the author has used in healthcare settings to engage staff members in outcomes improvement. Models created using these steps have proven effective in improving outcomes and saving millions of dollars because the process engages the entire healthcare team to provide input into (a) the design of measurement instruments, (b) interpretation of results, and (c) application of interventions, based on the data, to improve outcomes. Analysts and staff members build models of measurement that tell the story of the organization empirically, which makes the data not only actionable but relatable.

The Art and Science of Making Data Accessible

Data can and should read like a story. The presentation of data in healthcare should be interesting and engaging because it reflects empirically what people are experiencing operationally. It is the experience of this author that when data is presented as part of the employees' story, they love it. Everybody likes to talk about their experience, and when a data analyst is able to tell

The presentation of data in healthcare should be interesting and engaging because it reflects empirically what people are experiencing operationally.

Using Predictive Analytics to Improve Healthcare Outcomes, First Edition.
Edited by John W. Nelson, Jayne Felgen, and Mary Ann Hozak.
© 2021 John Wiley & Sons, Inc. Published 2021 by John Wiley & Sons, Inc.

The stories machines tell leave out the context, rendering their stories unrelatable.

them what they themselves are experiencing with the numbers to back it up, it places staff members at the edges of their seats.

With the advent of big data, machine learning, and artificial intelligence, we now too often turn one of our oldest, most cherished human traditions—storytelling—over to machines. The stories machines tell reveal patterns and relationships that staff members are familiar with, but they leave out the context, rendering their stories unrelatable. If your goal is to provide people with information they instantly recognize as accurate and relevant, your models must be specified to the people and contexts they presume to report on, and only then should they be examined empirically.

You are about to meet a 16-step process for how to tell a story, using data, that is not only interesting; it is actionable operationally. No two organizations are the same, and no organization stays the same over time. Thus, it is critical to evaluate whether data presented within an organization accurately captures the context and nuance of the organization at a point in time.

Admittedly, the idea of 16 steps may initially feel prohibitively complex. As you spend time looking at the process in terms of some practical examples, however, you will find that what I have provided is simply a template for examining and sorting data which you will find not only simple to use, but ultimately quite liberating.

If some of the content is foreign to you or seems beyond your reach, rest assured that someone on the team will know just what to do.

As you read through the steps, you are likely to intuit what role you would play and what roles you would not play, in this process. Some of the work described in the steps will be done by staff members closest to the work being analyzed, and some will be done by mathematicians, statisticians, programmers, and/or data analysts. If some of the content is unfamiliar to you or seems beyond your reach, rest assured that someone on the team will know just what to do.

Step 1: Identify the Variable of Interest

What problem or improvement opportunity is of concern and/or interest to your team? This could be an issue such as patient falls or the civility of employees, managers, or patients. This problem or improvement opportunity is referred to as the *variable of interest;* it will also serve as the focal point of the story your team will eventually be telling with data. In some settings, the variable of interest may be referred to as the outcome of interest.

Step 2: Identify the Things That Relate to the Variable of Interest: AKA, Predictor Variables

If your team was looking to improve an outcome related to falls, for example, you would want to examine anything that could predict, precede, or contribute to a fall. Assemble members of the care team and think together about what might lead to a fall, such as (a) a wet floor, (b) staff members with stature too small to be assisting patients with walking, (c) the patient taking a heart medication a little before the fall, and so on. As the discussion of everything that relates to your variable of interest continues, designate one person to write down all the things being mentioned, so the people brainstorming what relates to falls can focus solely on describing the experience and are not distracted by writing things down (Kahneman, 2011). Do not search far and wide for possible predictor variables or even think about the evidence from the literature at this point; just brainstorm and share. Variables from the literature can and should eventually supplement this list, but the focus in Step 2 is on the team's personal experience and subsequent hunches about variables that could affect the variable of interest.

Step 3: Organize the Predictor Variables by Similarity to Form a Structural Model

Have the team organize into groups all predictor variables that seem to be similar to one another. For now, you will simply separate them into columns or write them on separate sheets of paper. For example, one group of variables that may be found to predict falls may be "patient-related," such as the patient's age, level of mobility, the different diagnoses the patient is dealing with, and so on. These could all be listed in a construct under the heading "patient-related variables." You might also create a construct for "staff-related variables," such as "was walking with a staff member," "staff member's level of training in ambulating patients," "the staff member was new to this type of unit," and so on.

Step 4: Rank Predictor Variables Based on How Directly They Appear to Relate to the Variable of Interest

First, rank the individual variables within each construct, determining which of the variables within each construct appear to relate most directly to the variable of interest. These will be thought of as your most influential variables. Knowledge from the literature of what relates to the variable of interest is welcome at this point, but you should continue to give extra credence to the clinical experience of the team and what their personal hunches are regarding the relevance of each predictor variable. Once the predictor variables within each construct are ranked, then rank each overall construct based on how directly the entire grouping of predictor variables appears to relate to the variable of interest.

Step 5: Structure the Predictor Variables into a Model in Order to Visually Communicate Their Relationship to the Variable of Interest

It will eventually be important to get others on board who are interested in studying and improving the variable of interest. This can be done by noting the variable of interest in a circle in the middle of a blank page and then grouping all the predictor variables around the outcome variable so the visual looks a bit like the hub of a bike wheel with each predictor variable connected to the outcome variable by a spoke. Figure 1.1 is an example of a structural model that is ready to be converted to a measurement model.

For ease of understanding, items in the same construct would be the same color, and items in each color would then be arranged with those considered most influential positioned closest to the variable of interest. Selecting the most influential variables is important because you may decide you have only enough time or resources to address some of the variables. If this is the case, select those variables that are perceived to be the most influential.

A collection of three worksheets used by a neurosurgical nursing care unit for a study on workload, showing the progression from a full list of predictor variables to a workable model (Steps 2–5), can be found in Appendix A. These worksheets

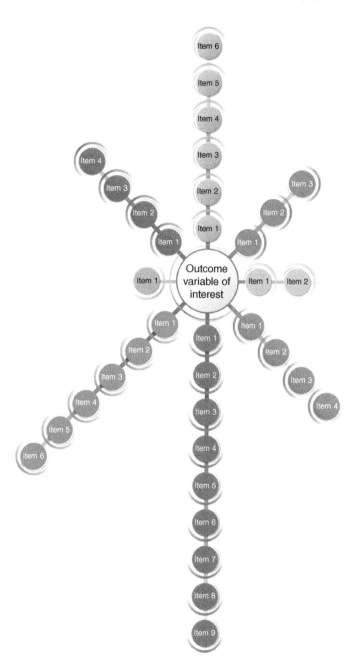

Figure 1.1 Variable of interest surrounded by constructs of predictor variables, arranged by rank.

visually represent the conversion of a structural model to a measurement model. While Figure 1.1 offers a visual representation of a model with multiple constructs, showing the variables arranged by rank, Appendix A shows a slightly different visual representation, which is the representation typically used by the author of this chapter. Readers are encouraged to try both methods of representing the constructs to see which is more useful in understanding and visually communicating the information.

Step 6: Evaluate if and/or Where Data on the Predictor Variables Is Already Being Collected (AKA, Data Discovery)

Investigate whether data on any of the predictor variables in your model is already being collected in current databases within your organization. Where you find that data is already being collected, you will use the existing data. You may find that data related to the variables of interest is being collected in more than one place, which will provide an opportunity for consolidation, making your data management process more efficient and standardized.

Step 7: Find Ways to Measure Predictor Variables Not Currently Being Measured

If influential variables are left out of the study, the model will remain mis-specified (wrong).

If there are important variables not being measured, it will be necessary to develop ways to measure them. If influential variables are left out of the study, the model will remain mis-specified (wrong).

Step 8: Select an Analytic Method

This work will likely be overseen by an in-house or consultant mathematician, statistician, or data analyst. Consider types of analytics beyond linear methods or qualitative descriptive methods, which are the most common methods currently used in healthcare. For example, if the dataset is very large and complex and it is not clear how to sort the predictor variables in a linear method such as regression analysis, try Pareto mathematics where outliers are examined. Pareto mathematics looks at the highs and lows in the dataset to create a profile of success factors. In his book *Where Medicine Went Wrong: Rediscovering the Path to Complexity*, Dr. Bruce West asserts that Gaussian mathematics throws out the extreme values despite the fact that these extreme values often provide the most valuable information because they provide a profile of the biggest successes and the biggest failures (2007).

Another type of analysis to consider is constructal theory, a method derived from physics, which is the study of "flow" (Bejan & Zane, 2012). If employees are able to talk about what makes their work flow or what makes their work pause, constructal theory will allow for their comments to be themed and addressed operationally. It is the experience of this author that if employees can talk about their workflow, and data can then be arranged for them in themes, the employees get excited about working on productivity because it is readily apparent to them that the overall aim is figuring out how to do more of what works well and less of what does not. Constructal theory makes productivity, or the lack of it, visible. Gaussian mathematics provides insight into linear processes, while analyses like Parato mathematics and constructal theory provide insight into more dynamic/complex and unknown processes, respectively. Pairing the analytic method with the variable of interest is important to achieving insight into the operations of work and associated outcomes.

The overall aim is figuring out how to do more of what works well and less of what does not.

For your most complex models, you may want to consider using a machine learning problem designed to let the computer tell you the rank order of predictor variables as they relate to your variable of interest. This is suggested on the condition that you never allow the machine to have the "final say." Machines function without regard for theory and context, so they are not able to tell a story capable of deeply resonating with the people whose work is being measured. It is tempting to be lazy and not do the work of carefully building models based on theory and rich in context that will result in data that makes people excited to take action. More on machine learning can be found in Chapter 6.

Step 9: Collect Retrospective Data Right Away and Take Action

If available, use retrospective data (data used for reporting what has already happened) to determine which predictor variables, of those already being measured, most closely relate to your variable of interest. Have a mathematician, statistician, or data analyst run a correlation table of all the predictor variables. Most organizations employ or contract with mathematicians, statisticians, and/or others who know how to run and read a correlation table using statistics software such as SPSS, SAS, or R. The mathematician, statistician, or analyst will be able to identify all the predictor variables found to have a relationship with the variable of interest and rank them in order of the strength of the relationship. (Many examples of these ranked predictor variables will appear in this book.) This will help start a meaningful conversation about what is being discovered while data on the remaining variables from the measurement model is still being collected manually. If actionable information is discovered during this step, operational changes can be implemented immediately. Further examination of the data will continue, but if what you have discovered by this point makes some opportunities for improvements apparent, do not wait to improve operations!

Step 10: Examine Data Before Analysis

Once the data is collected, it is tempting to proceed directly to the fun part and view the results. How do things relate to one another? What are the high and low scores? Viewing the results that answer questions like these is rewarding; however, prior to viewing any results (even in Step 9, where you may have gathered together some pretty compelling retrospective data), it is critical to ensure that all the data is correct. Have the statistician, analyst, or mathematician use statistics software, or even Excel, to examine the distribution of the scores and look for indications that there is missing data. If the distribution of the data for any predictor variable has a prominent leaning toward low or high scores, with a few outliers, the analyst will need to decide whether the outliers should be removed or whether the data should be weighted. If there are patterns of missing data (e.g. one item from a construct has many missing scores), then the group should discuss why the data is missing and what to do about it. Understanding the distribution and missing data will provide additional insight into the population being studied.

Reviewing the data for accuracy can also begin to give the team a feel for the "personality" of the data.

Reviewing the data for accuracy can also begin to give the team a feel for the "personality" of the data. This data all comes from people, but instead of a coherent conversation or observation of the people contained in the data, it is initially just data. As the data is *examined,* an understanding of the respondents will begin to emerge.

Step 11: Analyze the Data

For this step, it is important to engage a professional who is trained in analytics so the data can be interpreted accurately. There are several options in today's analytic software, such as SPSS, SAS or R, to aid in the examination of data flaws that are not obvious by merely looking at the dataset or its associated graphic representations. Your organization may have software, such as Tableau, which generates graphs automatically and is dependable for graphic visualization. In this step, it is important to engage an analyst or data scientist who can take advantage of the tools and tests contained in analytic software.

Step 12: Present Data to the People Who Work Directly With the Variable of Interest, and Get Their Interpretation of the Data

Ultimately, the most relevant and useful interpretation will be provided by those who live the clinical experience, because they can validate or refine the interpretation of the data. These people are typically clinicians, not analysts, however, so showing complex data in a format that is accessible to the employee *not* trained in analytics is important. When possible, use bubble graphs that provide a visual review of overall findings.

Figures 1.2 and 1.3 convey two ways to facilitate understanding of complex mathematics. Figure 1.2 provides a visual representation of the traditional graphic for results from a regression equation which examined three predictor variables related to reduction of central line-associated blood stream infection (CLABSI). The three predictors—location of the central line insertion, the RN assigned, and the phase of the project—have been tested to see the degree to which each of them explains the frequency of CLABSI.

Similarly, Figure 1.3 is a bubble graph which also shows how much of the outcome variable is explained by the same three predictor variables.

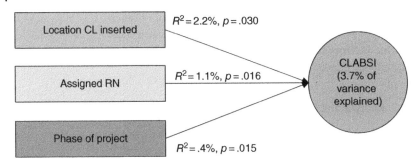

Figure 1.2 Explained variance of CLABSI, traditional graphic.

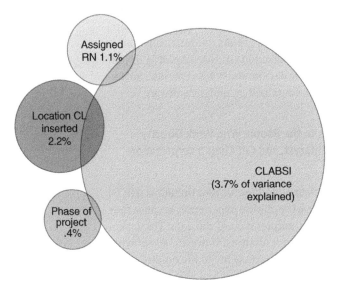

Figure 1.3 Explained variance of CLABSI, bubble graphic.

Both models reveal that the location of the central line insertion predicted 2.2% of CLABSI, the assigned RN predicted 1.1%, and the phase of the project predicted .4%. It has been the experience of this author that the visual representation in Figure 1.2 does not convey information as quickly as does the visual representation in Figure 1.3, in part because it relies too heavily on statistical symbols and equations to convey the information. In this book, however, you will see that most of the graphic representations of models and their results are expressed in traditional graphs, as many readers are likely to want more information than bubble graphs convey.

Analysts can provide a review of the results, but it is only the staff members who can provide validation or reflections that may suggest the need for secondary analysis to understand the data more deeply. When the data is being presented, pay attention to the listeners' responses. Even people who do not want to speak up may provide useful insight through nonverbal responses such as silence or even a shift in energy in the room. All of these cues can be informative. It is not uncommon for this author to pull listeners aside to discuss the nonverbal cues or silence that was observed. When encouraged to express themselves, these are often people from whom extremely valuable feedback is elicited.

Step 13: Respecify (Correct, Refine, and/or Expand) the Measurement Model

This is also the work of the data analyst, but it is done in close collaboration with staff members. This step includes refinement and possible expansion of the model to make it an even more sensitive model to detect predictors accurately. During presentation of the data to staff members, new variables will be identified, or variables that could not be measured in the first round but belong in the overall model will be addressed. Respecify the model to include anything that could not be included in the initial analysis or was identified in the interpretation as missing, and delete anything that was determined in the analysis to be unimportant. In one real-life example, we were measuring the performance of charge nurses as the variable of interest, and we had proposed that our three predictor variables were (a) demographics of the charge nurse, (b) attending the charge nurse program, and (c) the preceptor who trained the charge nurse into the role. The originally specified model looked like Figure 1.4.

After Model 1 was used to examine this variable of interest, and after the data was presented to unit managers, charge nurses, and staff members from the unit, those who attended the presentation reported that Model 1 was missing two influential predictor variables: (a) mentoring of the unit manager and (b) resources available on the job for charge nurses to execute their required role. These influential variables were added to a respecified model (Figure 1.5), and the study was

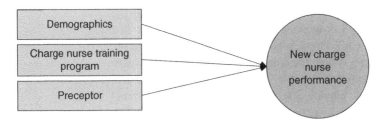

Figure 1.4 Model 1 to measure new charge nurse performance.

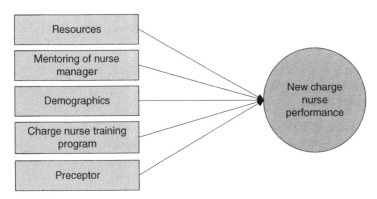

Figure 1.5 Model 2, respecified with new predictor variables to measure new charge nurse performance.

conducted again to see whether analysis of the respecified model could further explain what was influencing performance of charge nurses.

Note also that in structural models such as Figures 1.4 and 1.5, we have rectangles that look like they are representing one variable, when in many cases they represent multiple variables. For example, the rectangle labeled "Demographics" in both figures might be representing a dozen or so variables. These smaller, more compact models, which appear throughout this book, are called over-aggregated structural models. Remember when you see them that what looks like a model testing three or four variables is actually testing dozens of variables at the same time.

Step 14: Repeat Steps 2–13 if Explained Variance Declines

In nearly every instance, the data analyst will, along with staff, be repeating Steps 2–13. When initiating use of predictive analytics, conventional wisdom says that at least 50% of the variance should be explained using regression analysis, but it is the experience of this author that explained variance of 70–75% for a variable of interest can be achieved with a good fitting model, using 10 predictor variables or fewer, in a regression analysis.

As practice changes are implemented based on the information that emerges, variables from the initial model will no longer predict the variable of interest because the problem (or part of the problem) will have been solved by the practice changes. Traditionally, the analyst would then have to start over and develop a new model, but in this case, much of the work has already been done when developing the initial full model that is graphically depicted in Figure 1.1. As you return to Step 2, you will review the existing full model and rerun all the analytics to identify existing predictor variables that have now become an issue due to the new practice changes and/or identify new variables that relate to the variable of interest.

Step 15: Interface and Automate

Collecting data from a variety of software can take a lot of time, and it costs a lot of money for staff members to collect the data. These are compelling reasons to examine how technology can be used to automate the specified models developed to study the variable of interest. To examine outcomes in as close to real time as possible, interface the software and applications to one repository of data so the data can be examined as it comes in. Programs can be written to make the mathematical formula run every time one of the new variables comes into the dataset. A program can even be written for automatic respecification of the model as operations of clinical care improve. Manual respecification of a measurement model takes a lot of time, but if the program is set up to detect a fall in the explained variance for any variable, a program can be written that automatically reruns the correlations of all the variables in the model and then automatically builds a new model. Coefficients can be used to identify some specific aspects of how the newly added predictor variable is affecting the outcome variable. For example, if the variable of interest was CLABSI incidence, and the predictor variable is "central line type is causing infection," the

A program can be written for automatic respecification of the model as operations of clinical care improve.

coefficients can identify what *type* of central line is causing infections, what unit/department it is most likely to occur in, and/or other specifics from other predictor variables in the model.

Step 16: Write Predictive Mathematical Formulas to Proactively Manage the Variable of Interest

Over time, the analyses from models used to study how specific variables affect specific variables of interest will reveal trends that help us identify which variables pose the greatest and/or most immediate risks. Coding of the "variables of risk" into groups will allow you to use logistic regression or other procedures using odds ratios to automatically inform you of the probability that any given variable of risk (or group of risks) is actually causing an undesirable outcome. Real time analytics, made possible by the work you did in Step 15, will help you manage these risks *before* the undesirable outcome occurs. You might need to use more contemporary analytics, such as machine learning and simulation modeling for smaller samples. Machine learning and simulation modeling can also be used for testing reconfiguration of operations based on real-time risk. For example, with staff schedules, machine learning and/or simulation modeling can be used to test how staffing ratios of RNs to nursing assistants (and other skill mixes) affect safety.

Summary 1: The "Why"

These steps have been used by this author repeatedly to save millions of dollars related to healthcare outcomes such as reduction of patient falls, reducing central line associated blood stream infections, and decreasing length of stay. With the advancement of mathematics and technology, the widespread use of predictive modeling for proactive management of outcomes in healthcare environments is happening now and will only increase. However, the biggest caution this author has is to *not* let machines do the interpretation and validation of the data that must necessarily be done by the people who are actually carrying out the work. Big data and machine learning can now quickly scan large datasets for patterns in the data, but it is clear to this author that the data must always be examined by a trained analyst and interpreted and validated by the people closest to the work.

Summary 2: The Even Bigger "Why"

We know that organizations want to provide the highest possible quality, safety, patient experience, and financial performance, and, ultimately, that is why we do the hard (but surprisingly fun) work of predictive modeling for proactive management of these and other outcomes. However, this author must confess that the biggest satisfier of all is the level of engagement and sometimes pure delight that this work engenders in the people involved. What follows is a personal account illustrating how this work is consistently received.

A manager met me as I walked toward her unit to talk to her staff about their unit-specific results. She had in her hand the unit-specific report on their data, which described the state of affairs of job satisfaction. She exuberantly shared how pleased she was that she was so easily able to read her unit story and identify what aspects of the story were most important. She said they'd been calling the report their "bible of operations" because it provided a clear map for action planning. This "bible" was worn with use which informed me without words that they were entranced by their data and its application to operations. They could prioritize what needed to be addressed first, second, and so on, knowing that time and dollars would not be wasted on ill-conceived efforts. We measured data every 15 months, and by the time we measured again, they were eager to respond and see results generated by their unit-specific model that told their unique story. Her unit scores were among the highest in the 800-bed urban hospital, and the response rate for her job satisfaction survey increased every year until it was consistently at 100%. Staff knew that if they responded to the survey, their voices would be heard and acted upon. They all loved to not only hear their own story, derived from the data they provided, but to collaboratively work to make the next story even better.

When is the last time you read a report that was so meaningful, relevant, and helpful that it changed forever how you do your process improvement work?

When is the last time you read a report that was so meaningful, relevant, and helpful that it changed forever how you do your process improvement work? The biggest reason we do this work is because we have seen the power it has to improve the lives of patients, families, and everyone charged with improving quality, safety, the patient experience, and the financial performance of the organization.

Implications for the Future

Readers of this chapter are likely to relate to at least one of the 16 steps identified: administrators understand the outcomes; nurses understand the hunches; theorists understand the use of formal and informal theory; analysts understand the math; engineers, data scientists, programmers, and informaticists understand the movement from manual to automated data collection and reporting; and some of these people understand many or all of these 16 steps. Much of the guesswork is eliminated as hunches are tested mathematically before they are tested in practice. When healthcare organizations start using predictive analytics to improve outcomes, a big change happens in the care environment as better informed choices are made in how care is provided. Using the 16 steps described in this chapter will enable people in healthcare to move beyond managing negative clinical, financial, or operational outcomes, into a new paradigm of providing care. This move from reactive to proactive management of outcomes puts organizations light-years ahead of where they would otherwise be, while engaging teams in ways few of us have ever seen before.

When healthcare organizations start using predictive analytics to improve outcomes, a big change happens in the care environment as better informed choices are made in how care is provided.

2

Advancing a New Paradigm of Caring Theory

John W. Nelson and Jayne Felgen

A paradigm is a model or example of a way to view things. New paradigms seek to deliver new truths. This fifteenth century concept was applied to social sciences by Kuhn in an essay which proposed that new models and ways of thinking in science are always rebuffed by traditional views until a convincing argument is provided through careful assemblance of the new theory into a model (1962). Models for social and psychological constructs are built on observations and beliefs. Sometimes these observations and beliefs are discussed among scientists who then may conduct research, using models, to study the veracity and validity of these observations and beliefs.

The notion that "caring contributes to healing" is a paradigm held closely by people in various disciplines in healthcare, and it is the very core of the profession of nursing (Lazenby, 2017). This chapter discusses the paradigm that caring contributes to healing, and it describes a number of frameworks of care delivery that seek to operationalize and systematize caring behaviors and to promote, support, and nurture caring in all relationships in healthcare. Nursing has done an excellent job of generating caring theories and frameworks of care delivery, but a poor job of creating an argument rooted deeply enough in verifiable data and scientific rigor to convince the broader scientific community that caring contributes to healing. This book makes a scientific argument that caring does indeed contribute to healing. This chapter reviews the theories and

Using Predictive Analytics to Improve Healthcare Outcomes, First Edition.
Edited by John W. Nelson, Jayne Felgen, and Mary Ann Hozak.
© 2021 John Wiley & Sons, Inc. Published 2021 by John Wiley & Sons, Inc.

frameworks that have been tested throughout this book. Watson's Theory of Transpersonal Caring (2008a) is the predominant theory of caring science, and Relationship-Based Care® (Creative Health Care Management, 2017; Koloroutis, 2004) is the predominant framework of care. However, there are other caring theories and frameworks of care that are reviewed in this book as well.

A special note is made here about the framework of care called the Caring Behaviors Assurance System© (CBAS), because it is the framework of care for which there currently exists the most specified measurement instruments to capture the overall effectiveness of its implementation and outcomes and thus the argument that caring contributes to healing. Chapters 17 and 18 provide a thorough description of CBAS as well as a review of how its effectiveness has been successfully measured. It will not be reviewed in detail in this chapter on theory and frameworks because it is so thoroughly reviewed later and because earlier case studies in this book are taken from organizations using other frameworks of care.

Maturation of a Discipline

Nurses have revolutionized healthcare at least twice. Florence Nightingale's reformation of care delivery is the most well-known example. Prior to Nightingale, care delivery was provided by monks and nuns of religious traditions (Goodnow, 1929). Nightingale's method of care delivery in 1854 espoused the need for a clean environment and individualized care of the patient (Nightingale, 1959). By insisting on a clean environment, she ensured that germs were not spread as rampantly from patient to patient. In effect, her adoption and enforcement of the paradigm "environments of care must be clean" improved patient safety. Germs and how disease spread were not clearly understood in the mid-1800s, but scientific advancement is often discovered by accident (Kuhns, 1962). Nightingale's ability to document a decrease in mortality rate from 42.7% to 2.2% for soldiers in the Crimean war gained worldwide attention which resulted in her woman-only model of nursing becoming the standard of care across the globe (McDonald, 2001; Neuhauser, 2003).

A second occurrence in which a nurse revolutionized health-care is less well known. It was led by Agatha McGaw, a surgical nurse. It was McGaw who in the 1890s observed that patients often underwent surgery with great trauma and risk due to the methods used for anesthesia. It was her belief that this trauma and the associated deaths were due to the administration of too much anesthesia and/or the rapidity with which anesthesia was being administered. In response, she developed a method of administering anesthesia to patients by dropping the anesthesia medication into a cloth that was placed over the patient's nose and mouth. The anesthesia was taken in by the patient as the patient inhaled. The rate of dropping fresh anesthesia onto the cloth covering the patient's nose and mouth was adjusted based on the patient's rate of breathing, with fewer drops applied when the patient's breathing slowed, and more drops applied when the breathing rate increased. Using this method of administering anesthesia, she recorded giving anesthesia to 14,000 patients, none of whom died. Her observation of the patient, and adjusting treatment to the patient's response, resulted in the establishment of the profession of anesthesiology and a new paradigm of how anesthesia was administered, based on the patient's observed physical response (Koch, 1999; Pougiales, 1970).

Could the effectiveness of caring, as a proven way to hasten healing, be the third medical revolution put forth by nursing? Watson (a nurse) asserts that when nurses enact caring behaviors toward themselves and/or others, healing is facilitated (2008a). This is yet another paradigm that has come to be espoused by nurses around the world. While there was a surge of caring theories in the 1970s and 1980s, a recent search revealed that Watson's theory is the most referenced caring theory in literature, and it is the theory researched most extensively in this book. It is the conclusion of the editors and authors of this book and all of its chapters that the observable presence of caring behaviors is an important predictor of a satisfying and safe healthcare experience—both for patients and for providers in all roles and disciplines. All studies in this book took place in institutions pursuing implementation of caring behaviors using frameworks of care that espoused the importance of caring for both self and others. While outcomes indicate dramatic improvement due to the contributions of Nightingale and McGaw, such outcomes proving that caring for self and others improves patient health and nurse satisfaction are newer and fewer. This book contains research that goes a long way to solidify the paradigm "caring contributes to healing" as an evidence-based reality.

It is the intent of the many authors of this book to find the best theory or combination of theories to demonstrate not only that caring contributes to healing, but that the frameworks of care, through which caring behaviors are supported within systems of care, have inestimable value. We are at the point in this work where Benjamin Franklin was, just prior to his discovery of electricity. Prior to his famous encounter with the lightning bolt, there were several other theorists working on understanding electricity (Kuhn, 1962). It was not until Franklin considered the similarities of all the theories, and conducted a fascinating experiment with liquid electricity in a jar called a Layden jar, that Franklin's work resulted in what we now know as electricity as well as a profession for electricians to practice (Kuhn, 1962). Will one or more of the case studies in this book be the lightning bolt that shifts healthcare into a more complete and irrevocable understanding that caring contributes to healing?

Will one or more of the case studies in this book be the lightning bolt that shifts healthcare into a more complete and irrevocable understanding that caring contributes to healing?

Theory

Nursing works closely with medicine and other healthcare professions, but nurses are unique in that they are with the patient 24 hours a day and have the most consistent contact with the patient during care. This requires nurses to work within the systems of the organization and thus it is important to not only study nursing's many theories of caring, but how that caring is enabled or hindered by the social and technical aspects of the work environment.

Caring Theory

There are several theories of caring proposed in the profession of nursing, all asserting that caring is an important part of the healing process in care delivery. The most commonly cited theories are Watson's Theory of Human Caring (1979, 1985) and her more contemporary Theory of Transpersonal Caring (2008a). Several theorists have amended Watson's theory by adding concepts or combining it with other theories, including Duffy's Quality Caring Nursing Model (Edmundson, 2012), and Caroline Coates' theory (1997) which combines Watson's Theory of Transpersonal Caring with Bandura's concept of self-efficacy. Swanson's Theory of Caring Behaviors was derived from research Swanson conducted in obstetrics, which took into consideration the care experiences of both mothers and fathers (1999). The theories of both Watson and Swanson have been applied extensively in the studies reviewed in this book.

Sociotechnical Systems Theory

When studying the experience of the work of nurses, or any employee who works directly with patients, it is important to consider the ease or difficulty of operations for enacting care. The study of operations includes theories to help guide the development of research, methods to measure success, and interpretation and application of findings to achieve operational improvements. Sociotechnical systems (STS) theory proposes that both social and technical aspects of operations contribute to the experience of work (Trist & Bamforth, 1951; Trist & Emory, 2005). It has been established that while good equipment and resources are essential to carry out the technical aspects of the work, relationships contribute at least as much as the technical aspects of work to a productive and enjoyable work experience (Trist & Bamforth, 1951; Trist & Emory, 2005).

Misaligned, stressed, or missing relationships can impede productivity just as surely as poor equipment or resources can, which is consistent with concepts of pause and flow proposed in constructal theory (Bejan, 2019; Bejan & Zane, 2012). Bejan and Zane (2012) quote Michelangelo as saying, "Design is the root of all sciences" (p. 827). This ancient principle applies to the science of care delivery, both in how care is delivered by teams and how it is supported within systems. Constructal theory provides a theoretical framework with which to study the design of care, including the flow of work that results in enjoyment and productivity.

Frameworks of Care

It is important to distinguish a theory of caring from a framework of care. Typically a framework of care is based on one or more caring theories. What makes a framework different is that it includes a methodology for the theory to take hold in all of the structures, processes, policies, and people in an organization. A framework of care is designed to see to it that a theory of care, with its concomitant behaviors, becomes pervasive throughout a system. As mentioned, the two frameworks discussed in this book are Relationship-Based Care (RBC) and the Caring Behaviors Assurance System (CBAS).

Relationship-Based Care (RBC) is a framework of care with eight dimensions:

1) Patient and Family (in the center of everything)
2) Healing Cultures
3) Leadership
4) Teamwork
5) Interprofessional Practice
6) Care Delivery
7) System Design
8) Evidence

Like many frameworks of care, RBC engages staff councils in embedding its principles into the structures, processes, policies, and people in the organization. Most of the case studies in this book were carried out in organizations that practice RBC. Operationalization of the RBC model uses the powerful formula for change outlined in the book I_2E_2: *Leading Lasting Change* (Felgen, 2007) as it is key to a successful transformation. Appendix B offers a description of the I_2E_2 formula.

Another framework of care, which is reviewed at length in the international section of this book, is the Caring Behaviors Assurance System (CBAS). CBAS has six dimensions, which are based on the "7 Cs" derived from a 2010 paper published by the Scottish government, called *The Healthcare Quality Strategy for NHSScotland* [sic]. You will see that two of the 7 Cs are combined in the first dimension:

1) Care and Compassion
2) Communication
3) Collaboration
4) Clean Environment
5) Continuity of Care
6) Clinical Excellence

The chief difference between RBC and CBAS is that the CBAS framework seeks cultural change entirely through the implementation of behavioral changes for staff members. It is a framework, rather than a theory, because, like RBC, it engages staff members in embedding these behavioral changes in the structures, processes, policies, and people in the organization.

Unlike RBC, CBAS has an extensive method of measuring the degree to which the concepts of caring are taking hold in practice, employing three separate measures, including an assessment of how job satisfaction is being impacted by its implementation. Having a rigorous method of measurement specified for a framework of care helps advance and sustain the framework of care in several ways. It identifies (a) what specific components of the framework can be shown to relate to improved outcomes, (b) what components of the framework are critical for staff members to embrace in order to enact the framework of care, and (c) where the important components of the framework are working well within the organization and where they need additional support. The measurement process for CBAS accomplished all of these things. Implementation of CBAS in 18 hospitals in Scotland is reviewed in detail in Chapter 18.

There are several frameworks of care that are not included in this book but are reported in the literature. Sometimes frameworks are adopted in name only or amended dramatically for specific contexts. For example, the term patient-centered care is a general term proposed by the Institute of Medicine (IOM), which it defines as "providing care that is respectful of, and responsive to, individual preference, needs, and values, and ensuring that patient values guide all clinical decisions" (IOM, 2001, p. 5). This term is applied to a wide variety of models developed for specific organizations. Examples include the Cleveland Clinic's use of the IOM definition to create its own organization-specific patient-centered framework of care (Smith, 2018). Some companies have helped implement specific frameworks also termed patient-centered care. For example, Planetree has a number of methods to help organizations implement and become certified in a framework of patient-centered care (Planetree, 2020). Livestrong, a nonprofit organization which provides support for people with cancer, worked with the consulting firm Upstream to develop yet another framework of patient-centered care (Upstream, 2020). And finally, the Health Education of England (HEE), an agency within the National Health Service (NHS) of England, provides support to implement what it also calls patient-centered care (HEE, 2020).

Some theorists assert that Watson's Theory of Transpersonal Caring (2008a) is a framework of care when the processes of caring behaviors are taught with the intention that they be carried out within operations of care. While the implementation of any processes of caring in the absence of structures, processes, and policies to support the sustaining of the behaviors cannot be called a framework of care, the implementation of Watson's caring behaviors is an integral part of several frameworks of care delivery.

RBC's Four Decades of Wisdom

Later in this book, as case studies are presented in which people working in Relationship-Based Care cultures have had success in using predictive analytics to improve outcomes, they often credit the success of the project—or even the very existence of the project—to what they have learned while implementing RBC. Often, what they credit, however, is some of the long-honed wisdom baked into the process of implementing RBC that is not readily apparent in the dimensions of the model itself. You will see references to responsibility + authority + accountability (R+A+A), Primary Nursing, the importance of clarity, and more. In order to help you better understand those concepts when you meet them in your reading, here is a summary of each concept.

The Three Key Relationships in Relationship-Based Care

The most central tenet in Relationship-Based Care is that every relationship matters. Therefore, it is essential that all people in the organization tend to the quality of their relationships with themselves (self-awareness), with their colleagues, and with patients and families. Many discussions of these three relationships also focus on "care of" self, colleagues, and patients and families.

Relationship with Self/Care of Self

To stay healthy and be emotionally available for others, clinicians must pay attention to their own energy levels, be self-aware and mindful as they interact, and practice self-care for body, mind, and spirit.

Relationship with Colleagues/Care of Colleagues

Healthy interpersonal relationships between colleagues positively impact the patient experience. All team members must model mutual respect, trust, open and honest communication, and consistent, visible support of one another.

Relationship with Patients and Families/Care of Patients and Families

In RBC cultures, patients are seen, heard, and cared for as distinct individuals. Care and service are designed to prevent unnecessary suffering due to delays, physical or emotional discomfort, and lack of information about what is happening. The care delivery system of Primary Nursing, which is explained later in this chapter, is used because it is the system most supportive of the nurse–patient relationship (Manthey, 1980; Wessel & Manthey, 2015).

Several chapters in this book document studies seeking to understand the relationship between care of self and/or the care of the unit manager and nurse job satisfaction.

Responsibility+Authority+Accountability (R+A+A)

The theoretical framework known as R+A+A has appeared in nearly every book Creative Health Care Management (CHCM), the originators or RBC, has published (Felgen, 2007; Guanci & Medeiros, 2018; Koloroutis, 2004; Koloroutis & Abelson, 2017; Koloroutis, Felgen, Person, & Wessel, 2007; Manthey, 1980, 2007).

Any individual or group, whether council, task force, or committee, needs a clear scope of responsibility and a defined level of authority for decision making. R+A+A can also provide a mutually agreed upon system to assure reflection and review of the impact of everyone's efforts. Clarity around all three elements is necessary for success. They are defined as follows:

Responsibility

- Must include the clear articulation of expectations
- Always a two-way process: responsibility must be both allocated and accepted

Authority

- The right to act in areas in which one has been given and has accepted responsibility
- The level of authority must be appropriate for the responsibility

Creative Health Care Management uses four levels of authority to establish clear expectations for decision making:

- **Level 1**: Authority to collect information
- **Level 2**: Authority to collect information, assess, then make recommendations
- **Level 3**: Authority to collect information, assess, determine actions, pause to communicate and enhance, then act
- **Level 4**: Authority to assess and act, informing others after taking action

Accountability

- Ownership for the consequences of one's decision and actions
- Sets the stage for learning and directing future actions

When responsibility is understood and accepted and accountability measures are mutually agreed upon, people must always be given the authority necessary to be successful.

Several times in this book, a better understanding of R+A+A is identified as a key solution to management issues uncovered in studies on nurse job satisfaction.

Clarity of Self, Role, and System

In many of the case studies you will read in this book, clinicians wishing to improve practice look at the latent variable of "clarity." In various chapters, "clarity of self, role, and/or system" are shown to predict desirable outcomes. In order to facilitate greater understanding of those case studies, here is the "clarity of self, role and system" section of the "5Cs for Leading Lasting Change" (Felgen & Koloroutis, 2007).

Felgen and Koloroutis posit that any time courageous leaders in any roles at any level successfully challenge the status quo, five critical conditions—clarity, competence, confidence, collaboration, and commitment—are in place. What follows is a brief description of the condition and mindset of clarity:[1]

Clarity of Self

I am clear that I am an instrument of healing—that what I do or don't do matters. Either way, there is an impact. I am clear that I am responsible for my own growth and contribution to the team.

Clarity of Role

I know the broadest boundaries of my role and the roles of others, so I/we can optimize our individual and collective efforts for the team and the patients and families we serve.

Clarity of System

I embrace the mission, vision, and values of the organization. The operating functions behind the scenes can make or break our organization's ability to deliver the best care. Therefore, we must involve and empower first-line staff members to shape our systems to be reliable and effective for patients, families, and teams.

It is gratifying to see, in nearly every case study in this book, how many people in organizations across the world have resonated with this idea that we have long known to be so important.

Primary Nursing

Primary Nursing is the care delivery system embedded in the Relationship-Based Care model. In 1980, a little more than a decade after Primary Nursing's creation, Marie Manthey, one of its originators, wrote, "Primary Nursing is a delivery system for nursing at the [unit] level that facilitates professional nursing practice despite the bureaucratic nature of hospitals"

---

1 Felgen & Koloroutis, 2007.

(Manthey, 1980, p. 1). It was founded in the late 1960s by a group of highly committed nurses who were unsatisfied with the task-focus of team nursing and functional nursing. Neither of those systems of care delivery gave these nurses enough time to really know their patients, and they agreed that care was lacking when the patient was seen as the "reason for the tasks" rather than a person with wants, needs, thoughts, and concerns (Wessel & Manthey, 2015). In Primary Nursing, one professional nurse oversees the care of the patient from admission to discharge in inpatient settings (Person, 2004) and is assigned to the patient upon subsequent visits in outpatient settings (Wessel & Manthey, 2015).

Past measurement of Primary Nursing has included (a) assignment of the primary patient relationship at or soon after admission, (b) assessment of the continuity of care, (c) collaboration of the nurse with interprofessional colleagues involved with the patient's care, (d) care planning with the care team, (e) patient and family satisfaction, and (f) the relationship between patient and nurse (Nelson, 2001). The importance of Primary Nursing to nurse job satisfaction has been confirmed using factor analysis of the Healthcare Environment Survey (HES) in the United States (Nelson, Persky, et al. 2015), Scotland (Nelson & Cavanagh, 2017), and Turkey (Gozum, Nelson, Yildirim, & Kavla, 2021).

Several American studies in this book show that Primary Nursing is a chief satisfier of nurses. In studies around the world, it is often the case that while the term Primary Nursing is either not well known or has a negative association for nurses, the tenets of Primary Nursing (once the barrier of the term is removed) are consistently shown to contribute to nurse job satisfaction.

Summary

While caring is a traditionally embraced construct in nursing, it will require greater evidence to advance it as a pivotal part of a new paradigm in healthcare. Frameworks of care delivery are necessary to operationalize the paradigm that caring contributes to healing, and the frameworks of care that are explained in this chapter and studied in this book hold caring for self and others as central to effective practice.

If you think of clinical competence as being the combination of relational competence and technical competence—which implies that there is no clinical competence without relational competence (Koloroutis & Trout, 2012)—you will understand more fully how important it is to measure the effect that caring, clarity, and relationships have on healthcare outcomes. As you learn about the ways in which predictive analytics are being used to improve outcomes in healthcare, notice the vital role that the context of care plays every time. It is not just the technical aspects of care that are being measured now, because it is not just the technical aspects of care that matter. Caring, clarity, and relationships matter, so we are measuring them.

There is no clinical competence without relational competence.

3

Cultivating a Better Data Process for More Relevant Operational Insight

Mary Ann Hozak

An organization's performance improvement plan has traditionally been based on data that measures a compilation of (a) demographic information, (b) prevalence of outcomes such as use of restraints or pressure injuries, and (c) percentage of policy compliance such as: How many falls? How many appointments were canceled? Were patients happy or unhappy? What percentage of the form was completed? Although helpful as a starting point for identifying and quantifying quality indicators, the only thing these scores really show us is an abundance of data points moving up or down each month. They indicate whether a specific task is being performed well or poorly, but they do little to help us understand the big picture. Since "task performance" is an inadequate measure of professional practice, we need to rethink what we are measuring and how we are measuring it.

Since "task performance" is an inadequate measure of professional practice, we need to rethink what we are measuring and how we are measuring it.

Data collection itself is challenging. How do you decide what to collect or where to collect it, and how can you be sure it is collected the same way for every audit or that every auditor is auditing the same way? Then, once it is collected, what is the best way to present the data to help communicate with others what new realities you have come to understand?

Most often, data is collected manually, with some information technology assistance, which is a very arduous process. Once collected, however, the data is not always analyzed to find and correct any collection errors and outlier information before

Using Predictive Analytics to Improve Healthcare Outcomes, First Edition. Edited by John W. Nelson, Jayne Felgen, and Mary Ann Hozak. © 2021 John Wiley & Sons, Inc. Published 2021 by John Wiley & Sons, Inc.

it is distributed to people in the organization. This not-necessarily-valid performance improvement data is often then reported and discussed in committee meetings and staff council meetings. As you can imagine, the action plans created in response to this data can cause some serious problems. Using data without screening for error first is like putting a ship to sail before checking if the ship has holes in it.

If the data is not checked for accuracy, anxiety and frustration build for staff members and leaders alike as the organization's inconsistent data is then reported to Leapfrog, Hospital Compare, and the Magnet® Recognition program. The experience of anxiety for staff members and leaders is tied to the reporting of this data to regulatory and accreditation organizations, because these organizations report the hospital-level data to the public who then make choices about what hospital they will go to. The frustration is tied to the ongoing struggle with trying to understand the fluctuation in scores. This fluctuation in scores, which is also frustrating to manage, suggests that other variables affecting the outcomes have not been measured or were not measured correctly. Given this pattern of collecting, distributing, and acting on flawed data, with no sustainable improvement toward the goals of high reliability, patient safety, and full reimbursement opportunities, the question is: How can we get our hands on data that includes everyone on all units and points us toward the precise actions we can take to improve operations?

Taking on the Challenge

In our organization, the first step in getting out of the problematic data cycle was realizing that the quality committee members were clinicians and business administrators, not data analysts who can translate all those numbers into valid, reliable, widely consumable information. It had become clear that the money we were spending on the work of the quality committee was like spending three dollars on a bottle of wine, and then expecting to have a substantive conversation about the virtues of a truly fine wine. If we wanted to go beyond shuffling papers full of data to having a data process that truly informs, we needed to invest time and resources into experts in this field. It was clear that a more systematic and scientific data management structure and equally systematic and scientific processes were needed if our organization was to positively impact patient care, staff satisfaction, and financial outcomes. To develop the appropriate structures and processes necessary to maximize the reliability, validity, relevance, and consumability of the current data, executive leaders hired a data analyst trained in research methods, measurement models, and predictive analytics to assist the team. It should be noted that this PhD prepared data analyst had been a bedside nurse for 11 years and was familiar with our framework of care, Relationship-Based Care® (RBC) (Creative Health Care Management, 2017; Koloroutis, 2004), so he understood our context very well. We felt as if we had struck gold.

In our initial and subsequent Magnet® journeys, RBC had been chosen as the care delivery model by the organization because it aligned so thoroughly with our mission and vision. Through our work implementing RBC, we quickly discovered that staff members who have clarity about who they are, what their role is, and how the healthcare system works, were demonstrating better and more sustainable outcomes (Hozak & Brennan, 2012; Nelson & Felgen, 2015; Nelson, Nichols, & Wahl, 2017). Showing evidence of this rapid improvement, however, required a different approach, since "clarity of self, role, and system" were very unusual things to measure. It required administrators to trust that even though measuring caring and clarity did not directly measure costs or outcomes, a model that improves caring and clarity *does* make a financial impact in the long run, and it was therefore worth the time and investment to study these variables.

A model that improves caring and clarity does make a financial impact in the long run.

The quality committee partnered with the data analyst to measure clarity of self, role, and system, as we learned there was already a measurement instrument developed that could show us what positive outcomes clarity predicts (Felgen & Nelson, 2016). We used predictive analytics to study how clarity related to nurse job satisfaction, and how this impacted (a) caring as reported by the patient, (b) caring as reported by the care providers, (c) sleep quality of the patients, and (d) HCAHPS scores. Through this work, we discovered that our hunches about what we were implementing in RBC were right: increased clarity resulted in increased job satisfaction which in turn improved caring as reported by both patients and care providers (Nelson & Felgen, 2015). We also found that staff members who had higher job satisfaction also had patients who reported higher HCAHPS scores and better sleep during their hospital stay. We did strike gold!

Through our use of predictive analytics, we measured what was long thought unmeasurable, as we demonstrated that the concepts taught in RBC did matter.

Our study used predictive analytics to answer the questions and question the assumptions of staff members and leaders. The data provided at every step in the process became the vehicle for us to go deeper into conversations about the everyday operations of the units. We now had predictive analytics to help us discover some things that would allow us to refine operations with an extremely high likelihood that our efforts would be successful. The aspects of care we investigated further, using the data, addressed:

The data provided at every step in the process became the vehicle for us to go deeper into conversations about the everyday operations of the units.

1) The quality of relationships on the unit,
2) How staff members assessed their own care delivery and that of others,
3) Whether there was consistency in our care delivery,
4) Whether the mission and values of the organization were visible and tangible in our care delivery, and
5) How well the hospital's operational systems worked for delivering care to patients and families.

While these were nontraditional things to measure, we had confidence that by understanding them better, we could understand more clearly what predicted other outcomes. These variables were all front-of-mind because of our work with Relationship-Based Care. And because these variables came directly from the people closest to the work, the results of this study described the actual behaviors of staff members and leaders, providing highly relevant, actionable information and insights regarding the care environment.

The data analyst guided the staff members and leaders to discover the information *behind* the data by learning to ask questions starting with "how" and "why." The time we spent together further refining the questions, digging deeper, and examining hunches to answer these questions allowed for truly helpful stories to emerge from staff members and leaders about the variables we were considering. These stories began to reveal possible predictor variables related to all of the variables of interest we sought to change.

One example of a variable that is always of interest is fall rates. Traditionally, in our organization, when the falls data for a unit was trended over quarters, action plans for the unit have usually included reeducating the staff on two of our current fall reduction strategies—use of the "falling star" program and yellow socks—as well as consistently asking whether the patient's call bell was within reach. Admittedly, some very helpful action plans were created based on these inquiries. However, the data is now also looked at in terms of meaningful questions such as *"Why* do falls happen more often in the bathrooms at night?" *"How* is rounding conducted at night?" and *"How* do staff members interact with the patient during rounding?" When decisions about what to measure are made using a process that includes discussion of a wide and relevant array of real-world influences, your subsequent measurements will produce results that allow people to improve operations and design safer care delivery processes. These discussions, which included people from the staff councils and leadership quality councils, helped take the conversation deeper, from unit-based to broad systems of care. They encouraged examination of how workflow happens, consideration of the various staffing roles and skill mix, and the question we ask every time: "Can it be done better?"

"PSI RNs": A Significant Structural Change to Support Performance and Safety Improvement Initiatives and Gain More Operational Insight

In our organization, a significant change brought about by these deep dives was the development of a "staff RN quality charge nurse" position. The charge of these nurses was to focus solely on performance and safety improvement (PSI). The PSI RNs are staff nurses who monitor care on a particular unit or division and identify the factors that impact practice. As experienced RN staff members, they can identify gaps in workflow and resources, redundancy in workload, and the competency levels of both new and experienced staff members. The PSI RNs noted where and when staff members created workarounds for care delivery because operational pauses prevented people from getting their work done. It made sense to look at workflow and workload at the unit level, since as first-line workers, these PSI RNs were using their insight and experience to assess what was occurring operationally in patient care. The action plans that came from the data they provided could then focus directly on solving real-life problems in the care delivery system. The thorough data collection the PSI RNs made possible meant we would get more and better insights into our day-to-day operations.

It is largely staff-to-staff discussions, not manager-to-staff mandates, that create effective planning to meet the unit-specific goals of excellence and positive patient outcomes.

Once the data analyst and PSI RNs were in place, the organization made sure resources were available for them to do deep dives as a team. As each unit has its own culture and specific service line, the PSI RNs and the staff members of individual units came together to review the data reports from the analyst and discuss the "whys" and "hows." They came up with responsive interventions that were specifically designed for their patient population, since what works in a pediatric unit, for example, may not work in an intensive care unit. This grassroots approach made sense to the team and led to greater staff buy-in. It is largely staff-to-staff discussions, not manager-to-staff mandates, that create effective planning to meet the unit-specific goals of excellence and positive patient outcomes.

The PSI RNs provided context and clarity, helping managers and staff members see how policies, standards, regulations, and best practice come together to define quality care. This support for the nurse manager helped facilitate the change to a high-quality/high-reliability culture on the units, ensuring that each unit was delivering high-quality, evidenced-based care around the clock. This work can be overwhelming in organizations where there is a clinical staff of 150 nurses in large units, such as in the emergency department. The addition of the PSI RNs provided a unique mediating and educating role which served to support both staff members and managers in establishing standards of quality that were consistent with policy.

The work of these teams also provided a great example of "measuring care delivery" when it came time to write our Magnet® document. The Magnet methodology sets the outline an organization must use to describe how data is used to improve care delivery as well as organizational and professional growth, within the context of the unit culture. Our new quality structure and process proved to be a benefit when it came time to write the Magnet application, because the data management process and the PSI RN role were based on an understanding of the drivers of the data, innovative action plans that answered "why" and "how," and most importantly, they demonstrated sustained positive outcomes.

The Importance of Interdisciplinary Collaboration in Data Analysis

Nurses have always been a driving force in quality. Whether at the chief nurse officer's level or leading the organization's quality improvement department, nursing is charged with providing and ensuring excellent patient care. However, application of the quality data is interdisciplinary by nature, as each discipline's care delivery impacts the others. In our organization, after the addition of a data analyst and the PSI RNs, people in other disciplines observed that nursing had made significant sustainable improvements in care and outcomes. They started to ask for consultation from nursing in the proper use of predictive analytics and its application for operational refinement and outcomes improvement. Leaders in radiology asked nursing to teach their staff members how to examine the data and how to use the same brainstorming techniques the nursing staff had used to develop action plans.

A memorable example of interdisciplinary impact on outcomes was the presumably "cost-effective" decision to change the color of caps used for central lines. Although to the products and purchasing team it was just a color, and the yellow ones were less expensive than the green ones, they had no idea the impact the change in cap color would make on our infection control practices. After the cap change, the catheter-associated infections in the ICU began to rise, and it was not until the interdisciplinary team dove deeply into the discussion of line maintenance, and then finally down to cap use (an element

Although to the products and purchasing team it was just a color, and the yellow caps were less expensive than the green caps, they had no idea the impact the change in cap color would make on our infection control practices.

rarely thought of), that we discovered what had gone wrong. It was becoming clearer to people throughout the organization that everybody's input matters.

Key Success Factors

There are two key success factors that I would recommend to organizations pursuing the sort of quality improvement work outlined in this chapter. They are (a) hiring nurses with MSN and BSN degrees to work as PSI RNs, and (b) using recognition as an incentive for the good work done on quality improvement projects.

Education Levels of PSI RNs

Advanced practice nurses and clinical staff members with baccalaureate or master's-level education have been taught to be consistent in asking the "why" and "how" questions that need to be asked in the data analysis process. When the question arises: "Did we look at this specific factor?" these highly prepared clinicians tap into their knowledge of the subject and ask the necessary questions to get at meaningful, actionable data that can be used to improve care outcomes. Leaders need to advocate for nurses who are hired as PSI RNs and into other performance improvement positions, to be MSN and BSN level nurses, and they must support nurses at *all* levels of preparation in seeking continuing education in areas such as statistics, research, ethics, and practice standards. (Fun fact: Our PhD prepared data analyst was once an associate nurse.) This emphasis will promote sustainability and encourage a culture of innovation.

Recognition of Excellence

Staff recognition is also crucial to sustaining the changes that bring about improved outcomes. When a staff council makes a change that improves practice, giving thank-you notes, awards, and parties lets staff members know they are appreciated for the great work they have done. Be as specific as possible in your expressions of appreciation, focusing on specific improvements in practice and changes people have made that are meaningful to professional care. When staff members are recognized, they are motivated to keep working to improve things. They are proud of the outcomes they have achieved, and they love to know that leaders are also proud of them.

Summary

Healthcare is now very competitive, and the public often chooses healthcare based on online consumer advocacy resources. Improved outcomes help the organization score well with online resources such as Leapfrog and Hospital Compare, negotiate with payor sources, and even successfully achieve Magnet recognition. All of these financial benefits and awards are dependent upon having good and accurate data, but the true benefit and highest reward come from refinement of a caring environment for the patients and families where safety and quality come first.

As you dig into the case studies in this book, take with you the assurance that all of the work described in those case studies is within reach for you and your organization, too. If we can do it, you can do it.

4

Leadership for Improved Healthcare Outcomes

Linda Valentino and Mary Ann Hozak

All of the improved healthcare outcomes reported in this book can be traced back to a leader who understood that data must resonate with people. Data engages people when it is generated using measurement instruments specified to capture data that paints a picture of the organizational culture and associated operations. Highly specified data inspires action because it shows staff members what they are already doing well or need to fix to achieve improved operations and outcomes. When we, the authors of this chapter, share specific, recognizable, highly relevant data with staff members, they are most often at the edge of their seats. As they listen for the next piece of data, they anticipate the discussion because they already have ideas to improve operations. This chapter shares how leaders in healthcare organizations used data to make carefully targeted operational changes to improve outcomes.

Both of us have worked closely with a data analyst and have learned from him not only how to identify patterns within our data that reveal what is really going on operationally, but even more vitally, the importance of giving staff members access to data that really resonates for them. Once leaders from every level in the organization are working to tell their story more completely using data, and prioritizing their actions accordingly, widespread cultural and operational transformation begins.

Using Predictive Analytics to Improve Healthcare Outcomes, First Edition.
Edited by John W. Nelson, Jayne Felgen, and Mary Ann Hozak.
© 2021 John Wiley & Sons, Inc. Published 2021 by John Wiley & Sons, Inc.

Data as a Tool to Make the Invisible Visible

Data can help you identify leadership or teamwork issues such as incivility, and it can help bring factual realities into difficult conversations. What follows are two ways that data rather predictably reveals issues in organizations and units across the world, and one example of how the use of data can help facilitate difficult conversations.

Identifying a Bully

In the United States, the term "bully" refers to an employee who harasses or even abuses other coworkers (Lever, Dyball, Greenberg & Stevelink, 2019). Data derived from the Healthcare Environment Survey (HES), which measures nurse job satisfaction, behaves in a particular way when a bully is present. If satisfaction with coworker relationships is low and variables such as professional growth and satisfaction with patient care are high, it indicates that there is a bully on the unit. It is unknown how many times this pattern has been identified, but it has always proven to foretell the presence of a bully. (The data analyst can look at your data and tell you have a bully on your unit before he even visits.) It is also always the case that when this data is presented to the staff, the staff members all look at one another, knowing who the bully is, but not saying anything. The revelation of such data gives leaders greater urgency to address the issue. Bullies create chaos where they work, which causes relational distress for everyone. It is also the experience of the analyst that bullies typically work in pairs, with one person instigating the bullying behavior and the other coworker facilitating it. While it is often very difficult to move a bully out of an organization entirely, once the presence of a bully is revealed to all, action can usually be taken to start remedying the situation.

Managers Who Are "Buddies"

Another issue that can be identified by use of the HES is managers who are reluctant to enforce policies because they are too eager to be liked by their staff. Again, using the HES job satisfaction data, when the score for satisfaction with "relationship with unit manager" is high but all 10 of the other aspects of job satisfaction are noticeably lower, it is often because the unit manager has become "one of the gang" and is not tending to the work of actually leading the unit. In this "manager-as-buddy" scenario, it is also usually the case that the manager complains along with the employees and does not hold the staff accountable to follow policy, or understand their role, or accountability, or what their level of authority is in terms of patient care.

Something interesting (and initially alarming) shows up in the data when these "buddies" begin to evolve into good leaders. When leaders become clear in their own roles, they usually realize the importance of helping staff members become clear in *their* roles, which means, among other things, requiring them to follow policies. Not surprisingly, scores reflecting the staff's satisfaction with the manager will drop—sometimes precipitously— once the manager begins, after a period of having reliably been a "buddy," to really lead. What is fascinating, however, is to watch the 10 aspects of job satisfaction for the nurses under this person's supervision that were initially so low, steadily improve. If you understand the data, you realize pretty quickly that the falling score for satisfaction with the manager is not a reflection of the manager's poor performance, but actually a symptom of staff members adjusting to losing a manager who did not adhere to policy. It is common for the data analyst to have to spend some time encouraging managers who initially feel bad when their unit's score for satisfaction with "relationships with the unit manager" falls, especially after they have begun working so hard to be better managers. It can take 2–3 years for staff to adjust to the change when a manager shifts from one who lets people do mostly as they please to one who leads with role clarity and policy enforcement. If not for the presence of an analyst who understands the nuances of such data, the data would be disheartening, and nothing productive would likely come of it. It is the real-life story the data tells that is of inestimable value.

It is the real-life story the data tells that is of inestimable value.

Using Data to Bring Objectivity to Sensitive Topics

Sometimes data brings needed clarity to difficult subjects, such as tensions between doctors and nurses. It is not uncommon for nurses and doctors to have stressful relationships. As our data analyst colleague has seen time and again, this is often due to poor role clarity or poor system clarity in the independent, interdependent, and dependent functions of each role. The roles can overlap, and interdependence between nurses and doctors sets them up for tension if they lack role or system clarity in teamwork.

As you can imagine, simply announcing the reality that nurses in an organization or unit have said they are dissatisfied with their relationships with physicians would be counterproductive. When the data is understood in context, however, it can be used to facilitate some very helpful discussions and/or spur people to incorporate relationship-building opportunities into their action plans. Presenting the data as it relates to commensurate scores in role clarity or system clarity can help facilitate discussions on how people from each profession experience (or would like to experience) teamwork, or to identify where specific tensions exist in clinical care. Such discussions have helped reveal opportunities for resolution.

In an organization in Scotland, when the scores showed strained relationships between nurses and physicians, their solution was to hold retreats where nurses and physicians got together to discuss issues, both professional and personal, in a safe and nurturing atmosphere. It was a relatively small action that yielded noticeable improvement. And, of course, people do not take such actions in order to achieve a better score on a report; it is to have the care that comes with better relationships become standard in the organization.

Without effective leadership, change will not occur.

Without effective leadership, change will not occur. Leaders need to pay attention to what our data analyst colleague calls the "campfire experience"—the sitting-around-talking-excitedly aspect of data—and harness the energy of it. It is important to understand what is going on with the team's excitement. Once that excitement is in place, we can ask more specific questions: Should I be examining effective or ineffective leadership? What could we examine that would be of highest value to the team? What action does the data suggest we should take first?

Effective leadership can help staff members follow their excitement into a strategy.

Effective leadership can help staff members follow their excitement into a strategy.

Leaders Using Data for Inspiration: Story 1

Here is an example of how a unit manager used data to work with staff members to plan for change.

A primary goal of the nurse manager of a neuroscience unit was to create a sustainable Primary Nursing care delivery system as an antidote to the task-based nursing she saw on the unit. Seeing some hesitation in the group about where to begin, she had an idea for how to use data to get the staff council moving in the desired direction. She used call bell data obtained from the nurse call systems to get them started in the design of Primary Nursing on their unit.

The unit planned to build their Primary Nursing system utilizing RN pairs and partnerships.[1] Nurses would work in partnerships to help share select tasks such as routine nursing care, medication administration, responding to emergencies, answering call bell lights, and patient and family interactions. It was theorized that by working together as an intentional team for a select portion of the care, nurses and support staff members could serve their patients better.

The first thing they took on was the effort to decrease call bell response time, with consistent RN partners working together to figure out how to respond to patients quickly, with the goal of responding in under two minutes. As call bells rang within a range of the hallway, one of the RN partners would answer the call if the other was occupied with another responsibility. This initiative led to the nurses recognizing the complex aspects of their daily workload and enabled them to prioritize tasks differently. They used data to see what partnering strategies were working as they reduced the response time to call bells. These efforts contributed to improving staff satisfaction with both team relationships and patient care.

It was discovered that patient satisfaction was high only when the staff's satisfaction with collaborating with each other was also high. As the collaboration of the team increased and role clarity improved, the staff council saw a decline in call bell light response time, an increase in staff satisfaction with teamwork and patient care, and improvement in patient satisfaction scores.

Their ever-deepening understanding of how to use data even prompted the staff to inquire of the data scientist whether they could study the predictors of workload. The study that followed is described in detail in the article "Measuring Workload of Nurses on a Neurosurgical Care Unit" (Nelson, Valentino, et al., 2015).

As the collaboration of the team increased and role clarity improved, the staff council saw a decline in call bell light response time, an increase in staff satisfaction with teamwork and patient care, and improvement in patient satisfaction scores.

---

1 This new model of pairing and partnering assured the Licensed Vocational Nurse (LVN) and Certified Nursing Assistant (CNA) worked alongside and at the direction of the RN to decide who would perform what tasks. This was a more professional model than just doing tasks.

Leaders Using Data for Inspiration: Story 2

Here is an example of how a staff-level leader's enthusiasm for data led to some stellar, long-lasting outcomes.

A young nurse, in practice for about 3 years, volunteered to be one of the original members of the staff council in an organization implementing Relationship-Based Care® (Creative Health Care Management, 2017; Koloroutis, 2004). One of the major drivers of her interest was access to the clinical data and the opportunity to improve patient care with full access to outcomes of care. She had never had this sort of access before, and she was very curious about data and its implications for care delivery transformation. Her excitement at the opportunity to hear revealing new data about her unit, firsthand from the data scientist, was a large part of what drew her to take on the leadership responsibility of co-chairing the staff council. The data was presented and explained in such a way that it motivated her to create enhancements to the professional practice model on her unit while all of the information was fully available to the staff council and unit staff members. She recalls that the availability and transparency of the data led to building relationships among staff members and their manager, and inspired them to build a strong working team.

Her excitement at the opportunity to hear revealing new data about her unit, firsthand from the data scientist, was a large part of what drew her to take on the leadership responsibility of co-chairing the staff council.

Two major milestones stood out for her. First, she was struck by the power of the Primary Nurse–patient relationships on the unit. During her time co-chairing the council, this nurse cared for a chronically ill cancer patient who was about her age. As her Primary Nurse, she developed a relationship with this patient and was shocked and humbled when the patient requested—rather *insisted*—that she have her surgery on a day the Primary Nurse was working. This pattern of the patient coordinating her care with her oncologist and her Primary Nurse continued for months. Over time, patient and nurse worked together to be sure the Primary Nurse was working at the same times the patient came in for treatment. This event represented a milestone in the rollout of the Relationship-Based Care model. This nurse's actualization of autonomy as a Primary Nurse became a beacon to all of the staff councils and unit staff members.

This nurse's actualization of autonomy as a Primary Nurse became a beacon to all of the staff councils and unit staff members.

Secondly, as a co-chair of the staff council, this nurse engaged in using data to identify barriers to the implementation of Relationship-Based Care. As staff council members began to recognize patterns in the disruption of patient care, they requested permission from their nurse manager to start a support group for patients and families on their unit. The manager fully supported this idea and gave the nurses one hour per week of protected time in which to conduct a regular support group. The unit social worker, in the role of co-chair of the Patient-Family Advisory Council, also supported this effort. An interdisciplinary team met weekly with patients and families to listen to the ways in which the team could make care better for them and other patients. The nurse recalls that practical suggestions emerged from these meetings such as the need to have sandwiches stocked on the floor for patients at night and the need for family members to have more direct access to the Primary Nurse or associate nurse. The meetings produced ideas on which the practice council could take action to implement changes to enhance patient and family satisfaction. The unit-based Patient-Family Support Group, created by the staff council, grew to be the model for an institution-wide Patient-Family Advisory Council that is still in place today.

How Leaders Can Advance the Use of Predictive Analytics and Machine Learning

Thus far it has been reviewed how leadership at every level of an organization can help advance a framework of care such as RBC. What is most promising with all of the technology we now have available to us, however, is that once we set up systems of data collection to monitor change in a culture, we then have the ability to integrate that cultural data into more complex models of measurement. Access to these complex models of measurement moves the organization from mostly retrospective use of data to the proactive use of data to manage healthcare outcomes before they occur.

What if you could know the probability that a proposed change would be effective, before you put a lot of time and energy into implementing the change? What if you could know something about an initiative's effectiveness *in your organization* before you rolled it out to your staff? That is what predictive analytics and machine learning can offer you.

Many other professions use predictive analytics to learn the predictors of outcomes and then move on to machine learning and forecasting of risk. This became clear at the mathematics conference mentioned in the preface of this book. Dr. Nelson listened to researchers in logistics and mining where they used complex measurement models in simulations to study how changes to the process of work, using existing data, could be understood prior to making actual changes in the work process.

In healthcare, this same technology is available to leaders who are willing to harness the expertise of existing staff members, each possessing unique knowledge in different domains, to form teams capable of working in predictive analytics and machine learning. For example, programmers in most healthcare organizations are typically asked simply to program the data points necessary to track and manage outcomes. What most leaders in healthcare do not realize is if programmers could work closely, for even a short time, with analysts trained in structural and measurement models, many programmers could carry out most of the predictive analytics and machine learning operations that the top payed data scientists do. The challenge with leaving the programmer and analyst alone is that they almost always lack an understanding of the processes of clinical care. If a team of three could work together, a programmer, an analyst, and a clinician who also works in systems (e.g. a clinical leader), they could program their technology to tell the relevant stories of healthcare. This team could design measurement instruments to capture the extent to which a cultural change is taking place and even to predict outcomes.

Dr. Nelson has found that his 11 years as a bedside nurse, his love of using mathematics to tell a story, and his years of work in software development have served him well in "seeing the stories" as he views data. It is these "stories" that make complex measurement models possible. Anything can be measured if its story can be articulated. The challenge is getting people together who can not only hear the story but can then assemble models and use technology to predict what will happen next.

Understanding an Organization's "Personality" Through Data Analysis

Spending time with our data analyst helped us understand many new things. For example, we now realize that collecting data is like interviewing hundreds of people. The interviews are carefully examined in order to understand the significance of the varied responses from people in different areas of care, people from different disciplines, and people from specific demographic groups in the organization. We learned that analyzing data is like carefully listening to what each person is saying. You can hear what individuals and groups are saying, and you can learn something about who is thriving and who is struggling. This can be an enjoyable event, and it can also be stressful when you notice the sometimes vast differences in experience and outcomes, even within the same organization, sometimes even within the same unit. Each organization or unit has an overall culture, almost like a personality, where various behaviors are appreciated and others are tolerated, sometimes to the detriment of everyone involved. "Listening" to the data makes possible the retelling of what was learned about the culture and experience of the people. In the hands of a skilled analyst, data can tell a heart-felt, and sometimes a heart-breaking, story.

Data can tell a heart-felt, and sometimes a heart-breaking, story.

Having valid data about the effectiveness of any dimensions of a framework of care delivery enables the people closest to the work to engage in action planning, and it also makes it possible to propose predictive analytics studies which would allow people to move beyond real-time use of the data into forecasting the likely outcomes of proposed changes. Organizational leaders and data managers can work together to tell the story of cultural change, prompted by the implementation of a framework of care, and use data to guide the change in the desired direction.

As you read each chapter in this book, we invite you to imagine that all or part of the story being told is happening in your organization, and how you and your colleagues might use this sort of information as you engage in the hard but joyful work of making your organization the best it can be.

Section Two
Analytics in Action

5

Using Predictive Analytics to Reduce Patient Falls

Tara Nichols, Lance Podsiad, Josephine Sclafani Wahl, and John W. Nelson

When a 350-bed hospital initiated Relationship-Based Care® (RBC) (Creative Health Care Management, 2017; Koloroutis, 2004) in their organization, they were already working on reducing falls; so when a team from the organization attended the RBC leader practicum, they had the issue of falls on their minds. After attending the practicum, this group of nurses had renewed concerns about the ways in which their current practice diverged from the most important tenet of Relationship-Based Care—the primacy of nurse–patient relationships—through its use of sitters to reduce patient falls. Sitters are never nurses, so they worried that the use of sitters represented nurses' abdication of responsibility for their patients.

These nurses and other leaders came to realize that the use of sitters was a symptom of a relationship problem among coworkers and between nurses and patients. Possible overuse of the sitter program came about because physicians had witnessed so many falls on the nursing care floors that they began to order sitters for patients quite frequently, specifically for patients who were combative and/or could not communicate, or who had a history of delirium and were at risk for becoming combative. The physicians' decision to use a sitter was based solely on risk, without considering how the relationship between patient and nurse might be adversely affected. Physicians from the emergency department and rounding physicians on the floors would all use this decision-making process when ordering a sitter to sit with the patient. This widespread use of sitters was also very expensive. For

Using Predictive Analytics to Improve Healthcare Outcomes, First Edition.
Edited by John W. Nelson, Jayne Felgen, and Mary Ann Hozak.
© 2021 John Wiley & Sons, Inc. Published 2021 by John Wiley & Sons, Inc.

example, a 600-bed hospital reported a monthly cost of sitters to be over $18,000 with some hospitals reporting annually spending $500,000–$2,000,000 (Davis, Kutash, & Whyte, 2017).

Not only did nurses on the floor feel insulted that they were not trusted to keep their patients safe, they also believed that overuse of the sitters was the *reason* for patients falling, because sitters did not have the clinical knowledge to assess the patients' needs related to fall risk. Still, their use of sitters allowed nurses to take a bigger patient load, so it was a while before the downside of using sitters became apparent.

After returning from the RBC leader practicum, the group of leaders, which included the chief nursing officer, taught RBC concepts in the organization's Leading for Accountability program. In this program, they learned about being accountable for relationships. As a result, they did three things to improve relationships, decrease falls, and save money:

1) All leaders in nursing services were educated in the concepts of Leading for Accountability. In this class they taught the RBC concept of responsibility + accountability + authority (R+A+A) which was described in Chapter 2 of this book.

2) They changed the decision-making process from being "sitter driven" to being safety driven. The decision whether to use a sitter was made collaboratively between the emergency nurse and floor nurse. If the patient had a history of delirium, past falls, or was unable to communicate and somewhat agitated, a safety-driven plan of care was initiated before any consideration was given to whether to use a sitter. For example, if the patient could not communicate and seemed combative but also had not urinated for a long period of time, a bladder scan would be performed to assess whether agitation might be from a physical need the patient could not communicate. Or if the patient had a history of delirium, the nurse would use interventions within the plan of care to help with orientation or call in a family member to help keep the patient oriented.

3) If the patient did need a sitter, the nurse would record in the plan of care the justification for a sitter, work with the physician to get an order for a sitter to sit at the bedside, and then educate the sitter on who the patient is as a person and what matters most to the patient and family, as well as educating

the sitter on the patient's plan of care related to fall risk. This helped sitters have clarity of role, take accountability for their role in keeping the patient safe, and establish a relationship with the patient.

By the end of the fourth month, sitter use was way down, and relationships were again restored. The cost savings every month from the reduced use of sitters ranged from $20,000 to $40,000.

While the reduced use of sitters enhanced the nurse–patient relationship and saved money, falls did not decline. This reality helped the nurses understand that they needed more data. The clinical nurse specialist, who was also the organization's RBC implementation lead, discussed this entire falls program with a data analyst who was at the organization assessing the state of caring and job satisfaction as it related to Relationship-Based Care. He said, "It sounds like you should study this issue using predictive analytics and start by building a model of measurement to study falls within the context of RBC." A plan to do so was undertaken.

While the reduced use of sitters enhanced the nurse–patient relationship and saved money, falls did not decline.

Predictors of Falls, Specified in Model 1

The falls committee from the general patient-care units (GPUs) took this project on and was eager to work with one specific GPU to study falls because they were struggling with a high fall rate. An invitation was made to the critical care units to participate as well, but they declined. The staff council from the GPU with the highest fall rate worked with the falls committee to develop a model of measurement that would adequately represent the predictors of falls in their care context. They worked with the data analyst to build the model in Figure 5.1, using several resources which included:

1) Variables identified in an international study on falls in which they had recently participated,
2) Data that was already being collected about falls in their electronic medical record (EMR),
3) Variables identified by the hospital's falls committee, and
4) Variables they had hunches about based on their own clinical practice and experience.

The specified model developed for this team included 53 variables they identified as candidates for data collection and subsequent analysis (see Figure 5.1). For variables from the falls models that were not in the EMR, they created an Excel spreadsheet and collected that data manually.

This group collected data on patients who had fallen and patients who had not fallen, as they understood that in order to study why one group of patients on their unit fell, they needed the same data for a group of patients who did not fall. To collect this data, they proposed that each time a patient fell, data on that patient would be collected, and each time that happened, data would also be collected on another randomly selected patient who had not fallen.

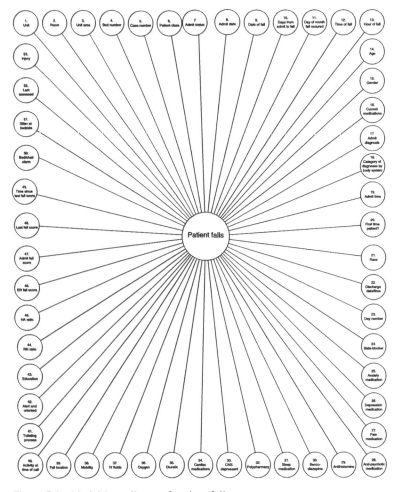

Figure 5.1   Model 1 predictors of patient falls.

After a power analysis was conducted to establish the number of patients needed in the study, it was determined that at least 300 patients total (with roughly equal groups, 150 who fell and 150 who did not fall) would be needed in a regression analysis if we were to understand what factors related most strongly to why patients fell.

Ultimately, data was collected from 163 patients who did not experience a fall and 170 patients who experienced a fall over a 9-month period. All patient ID numbers were given a code in the Excel file to protect patient privacy during the data collection period. Most of the patients experienced only one fall, but two patients fell twice. Each predictor variable was examined using regression analysis of the 170 fall patients and 163 non-fall patients. Results of each regression analysis revealed the predictor variables that related to falls:

- Diagnosis predicted 12.5% of variance of falls ($p = <.001$). Neuro, prior fall, and weakness (in that order) were most predictive of increasing falls.
- Mobility predicted 11.5% of variance of falls ($p = <.001$). Patients who were reported to need only minimal assist with mobility were at greatest risk to experience a fall.
- Last fall risk score predicted 9.9% of variance of falls ($p = <.001$). The higher the most recent fall score, the more likely the patient would fall.
- Area of the unit predicted 7.9% of variance of falls ($p = .001$). Patients on the south most portion of the unit were more likely to fall.
- Toileting process predicted 5.9% of variance of falls ($p = .001$). Use of commode and/or urinal were most predictive of more falls.
- Cognition (A&O) predicted 4.5% of variance of falls ($p = <.001$). A score lower than 3.5 on a 4-point scale was most predictive of more falls.
- Admitting fall risk score predicted 4.5% of variance of falls ($p = <.001$).
- Bed/Chair alarm predicted 4.4% of variance of falls ($p = .001$). Having a bed/chair alarm was found to be most predictive of more falls.
- Diuretic predicted 1.9% of falls ($p = .016$). Not having a diuretic was predictive of more falls.
- Educating the patient on their individual risk of fall predicted 1.3% of falls ($p = .039$). Patients who were educated on personal fall risk were less likely to fall.

The final regression analysis, studying all 10 variables at the same time, revealed 3 variables to have the most influence on why patients fell in this organization. In this analysis, diagnosis predicted 19.3% of why patients fell, mobility predicted 11.2%, and the patient's last fall risk score predicted 4.5%. Combined, these three variables predicted 35.0% of why patients fell. These findings are noted in Figure 5.2.

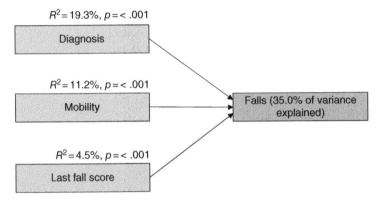

Figure 5.2   Model 1, examining what variables predict falls.

Lessons Learned from This Study

Examination and application of the findings, as they relate to both correlations of variables with falls and the final regression model of falls, were initially embraced more directly by the general patient-care units (GPUs) than they were by the critical care and intensive care units (ICUs). However, the findings were eventually embraced more widely because there had recently been a concerning upsurge in the incidence of patient falls. Staff on one of the ICU step-down units reported a peak of 19 falls in 1 month before the initial fall study on their 40-bed unit. Having so many falls was demoralizing and stressful for the staff members and leaders of the unit.

Findings of the study which were then used to guide operational changes and subsequently helped most to reduce falls are presented next.

Lesson 1: Age Was a Factor, But in an Unexpected Way

Viewing the correlates of the falls helped clinical care staff members think differently about falls. There had been a bias to think of elderly people as being at risk when advanced age did not actually factor into the correlations with falls. Looking at the patients who fell, there were 45-year-old men who needed to be up with minimal assist, but since the patient was not seen as elderly by the staff, the same level of caution was not provided. Despite anything the literature and the staffs' hunches would have suggested, the demographics of 45-year-old men were falling with greater frequency than anyone else. Having data the staff members could relate to operationally helped them think about patients from the perspective of the data.

Despite anything the literature and the staffs' hunches would have suggested, the demographics of 45-year-old men were falling with greater frequency than anyone else.

Lesson 2: Patients Resist Assistance When They Underestimate Their Own Fall Risks

When staff members treated patients with increased caution for a fall, it was not unusual for a patient to report feeling like they were being treated like they were helpless. This was due to the patient also having a bias about not feeling at risk for a fall. It was interesting to find that patients, like staff members, were biased about their own fall risk. Patients who felt they were being treated as helpless were surprised to learn of their own fall risk factors. After staff members educated these patients, the patients became more cautious in walking or toileting independently. Both staff members and patients were working with biases, and both benefited from educating patients on their individual fall risks.

Both staff members and patients were working with biases, and both benefited from educating patients on their individual fall risks.

Lesson 3: A Blame-Free Environment Accelerates Fall Reduction Efforts

Using the patient care council, which includes both staff members and management, was helpful in establishing organizational support for fall prevention. Leaders created a blame-free environment in reviewing causes for falls, which led to more transparent dialog and honest interpretation of study findings.

Lesson 4: Clarity Around R+A+A Is Essential

Staff members studied the concept of responsibility + authority + accountability (R+A+A) (Koloroutis, 2004; Koloroutis & Abelson, 2017). It was vital that staff members understood their responsibility and accountability in reducing falls and that they were given authority to act on interventions based on data.

Lesson 5: Informal Conversation Was the Biggest Factor in Spreading the Word About the Study

The findings were disseminated in a staff newsletter and in huddles, but the most helpful factor was how the data changed the conversation about falls in an informal way among clinical staff members. This informal conversation was most impactful among staff members who were cross-trained across GPUs and helped educate others about the data and interventions regarding reduction of falls. The GPU staff members were very excited about the decline in falls as it improved their confidence in providing care and decreased stress from feeling bad about patients being harmed while in their care.

Lesson 6: Complete and Accurate Data Are Essential to Any Study

This study would not have been possible without the data collector in the organization being meticulous about the accuracy and completeness of the data collected for the study. The completeness and accuracy of data helped increase the power and precision of the data which in turn made the findings resonate with the staff. This resonance helped staff members believe that all of the data was accurate and could be relied upon as we discussed changes in operations and policy.

The completeness and accuracy of data helped increase the power and precision of the data which in turn made the findings resonate with the staff.

Respecifying the Model

After a discussion of findings from Model 1, the model was respecified to include all of the variables from Model 1 that had a statistically significant relationship to falls, as well as those variables that were subsequently identified as missing from Model 1. This follow-up study included 237 patients who were studied for a 6-month period, using the same methods for data collection as before. This number was below our desired 300 patients, but the power parameters were considered adequate. Results from Model 2 are found in Figure 5.3.

In this next round of testing, because of practice changes implemented after the study of Model 1, all three of the variables from Model 1 either decreased in how much variance they explained for falls or disappeared altogether. In Model 2, diagnosis explained 8.6% of the variance of why people fell, which was

Because of practice changes implemented after the study of Model 1, all three of the variables from Model 1 either decreased in how much variance they explained for falls or disappeared altogether.

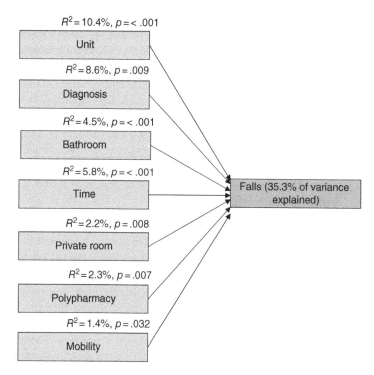

Figure 5.3 Model 2, examining what variables predict falls.

down from 19.3% in Model 1. This was partially attributed to staff members thinking differently about falls. Because of their earlier findings and subsequent education efforts, they realized that patients who were on medications that caused dizziness needed to be educated on risk for fall and be treated with more caution.

Mobility also declined as a predictor of falls from Model 1 to Model 2, declining from 11.2% to 1.4%. Staff members from the GPUs reported that mobility was no longer contributing to falls because staff members became aware of their own biases (e.g. assuming that advanced age was a strong predictor of falls) and began to instead use the newly acquired data on falls to guide practice. Due to patients who were up with minimal assist being at the greatest risk for a fall in Model 1, staff members became very conscious of falls when caring for patients with this level of mobility, no matter what age the patient was.

Last fall score predicted 4.5% of the variance of falls in Model 1, but it disappeared as a predictor in Model 2. This was because the organization changed from using the 4-point fall risk score that made nearly everybody seem to be at risk to the more sensitive 27-point fall score. Examination of the falls by fall risk score had revealed that falls increased in patients with scores between 13 and 17. Identifying that falls increased at the score of 13 helped staff members create an intervention to increase fall risk surveillance. Increased surveillance included reporting the score of 13 or more in the nurse-to-nurse change of shift report and clinical care rounds and integrating it into the patient' clinical plan of care document, which was used by the clinical care team.

New risks found in Model 2, that had not been found in Model 1, included the following:

- Unit predicted 10.4% of the variance of falls ($p = <.005$). Patients who were cared for in either of the two intensive care units increased the risk of falling. (Remember, the ICU staff were not initially as engaged with use of the data to decrease falls, when contrasted to the GPUs. The dramatic improvement of the GPUs decreased the risk so much that they became the units where falls were least likely to occur.)

- Time of day predicted 5.8% of the variance of falls ($p = <.005$). Later in the day was a greater risk for falling than earlier in the day.

- Being in the bathroom predicted 4.5% of the variance of falls ($p = <.005$). Patients were at most risk for falling in the bathroom, versus the main part of the room or hallway.

- Polycopharmacy predicted 2.3% of the variance of falls ($p = <.005$). Being on multiple medications increased the risk of a fall.

- Being in a private room predicted 2.2% of falls ($p = <.005$). Being in a private room increased risk of a fall.

The dramatic improvement of the GPUs decreased the risk so much that they became the units where falls were least likely to occur.

Summary

This organization sought to decrease falls from four falls per 1,000 patient days to less than one fall per 1,000 patient days, which was below the national benchmark recognized in the literature. And remember the ICU step-down unit that had been so demoralized after having 19 falls in 1 month? They took their total falls down to zero in the first month after the study and have consistently kept their falls below benchmark. The team leading this study and all of the people participating in it used relational concepts from Relationship-Based Care, educated the patients on their fall risk, and worked with a shared governance team to decrease falls. The use of these three strategies, coupled with the use of predictive analytics, resulted in a sustained decrease in falls and a savings to the hospital of $1.6 million in 1 year. (To learn how these cost savings were calculated, see Appendix C.) Once the data was collected and presented to the staff, it took only three months to implement the resulting practice change recommendations and to see fall rates plummet.

6

Using the Profile of Caring® to Improve Safety Outcomes
John W. Nelson and Kenneth Oja

This chapter has two aims: (a) to review the use of automated predictive models (machine learning) to improve outcomes and (b) to show how models that include the Profile of Caring® will improve safety outcomes by providing an understanding of what specific clinician behaviors, beliefs, and conditions lead to safer outcomes. This understanding will make it possible to proactively hire and train staff for safety.

The Profile of Caring

Before moving into a discussion about machine learning, we would like to give you a brief orientation to the Profile of Caring. As you encounter what we have found in the literature on outcomes throughout this book, take note of the prevalence of the collection and examination of patient data and administrative data in predictive models, almost always in the complete absence of the examination of the beliefs, behaviors, attitudes, or even skill levels of the people providing care. It is our hunch that the beliefs, behaviors, and attitudes of staff members are at least as impactful as patient and administrative data to the health outcomes of our patients. You can have the most state-of-the-art equipment in the world, but in the hands of a poorly trained, burned out, or unclear clinical staff member, you are not going to get a state-of-the-art outcome.

Using Predictive Analytics to Improve Healthcare Outcomes, First Edition.
Edited by John W. Nelson, Jayne Felgen, and Mary Ann Hozak.
© 2021 John Wiley & Sons, Inc. Published 2021 by John Wiley & Sons, Inc.

The Profile of Caring is a trademarked set of variables used to describe nurses who are experiencing (a) clarity of role, (b) clarity of system, (c) caring for self, (d) the caring of the unit manager, and (e) job satisfaction. The authors of this study believe that when clinicians in all disciplines fit the profile described by these five variables, care in all environments is improved.

The authors of this study have discovered through their research in patient care, and specifically in predictive analytics studies, that there is little research (we found none) in how the profile of staff members relates to outcomes. Are clinicians who have high report in each of the five variables in the Profile of Caring more likely to think more clearly and compassionately about the patients they care for? It is proposed in Watson's Theory of Transpersonal Caring (2008a) that when caring occurs, healing is enhanced. We would like to study this theory, and the Profile of Caring provides us, and anyone doing predictive analytics work in healthcare, with a set of variables with which to do so. The initial international study in which the five variables of the Profile of Caring were determined is detailed in Chapter 16 of this book.

Machine Learning

Utilization of machine learning (ML) and/or working with experts in ML requires, at minimum, a basic understanding of what ML is and the opportunities ML provides for making decisions to improve outcomes in clinical care. A basic understanding will help researchers[1] and clinicians decide whether they want to use their machines to maximize efficiency, test theory, leverage existing evidence, and/or be most accurate in their predictions of outcomes and refinement of operations.

The relationship between researchers and machines can be compared to that of a teacher and student in a statistics class. We will use the teacher–student analogy to help explain the

---

1 When the word "researcher" is used in this chapter, it is to be assumed that the researcher is always working with staff and clinical leaders and never an island of one.

four basic types of machine learning problems: supervised, unsupervised, semi-supervised, and reinforcement.

Supervised Problems

The teacher gives a student a dataset of patient clinical data and informs the student which variables to report on as the variables relate to falls. The information provided to the student is based on the teacher's theory and knowledge about falls. This is a supervised problem that has been assigned to the student. In a machine learning scenario, this would correlate to the researchers programming the machine to examine only certain variables and to analyze how those specific variables relate to falls, without looking for or analyzing other data.

Unsupervised Problems

The teacher gives the student the same dataset of patient clinical data but this time does not have knowledge or theory of the reason patients are falling and thus asks the student to examine the data and identify all possible relationships among the variables in the dataset. In this unsupervised problem, the student informs the teacher. In a machine learning scenario, this would correlate to the researchers programming the machine to analyze all data, examine all variables, identify all relationships, and report all information related to falls. Because machine learning algorithms build mathematical models without being explicitly programmed to do so, a machine working on an unsupervised problem informs the researchers about patterns, correlations, rankings, and other relationships, well beyond what the researchers know to ask about on their own.

Semi-supervised Problems

The teacher gives the student the same dataset of patient clinical data and has some knowledge about why patients fall, but there are several variables that are not understood in relationship to falls. Thus, the teacher informs the student which variables will relate to falls, but also asks the student to look for patterns among the other variables. In this semi-supervised problem, both the teacher and the student provide information. In a machine learning scenario, this would correlate to the researchers programming the machine to examine specific variables and analyze how they relate to falls, while also analyzing all other data, identifying and examining any additional variables, and identifying all possible patterns, correlations, rankings, and relationships.

Not all data scientists recognize semi-supervised problems for use in machine learning. However, the authors of this chapter believe it is important to include them here in order to help those new to ML to round out their knowledge of how the researcher and machine can work together.

Reinforcement Problems

A reinforcement problem structure is similar to that of an unsupervised problem, but it goes one step further with the teacher using a "reward system" to advance the learning of the student. Because the teacher knows the desired end result, the teacher can set an algorithm to reward students when they select the proper path toward the solution and "punish" students when they select the incorrect path toward the solution. The student is able to find the correct path by developing knowledge about patterns and relationships in the data. In the relationship between the researcher and the machine, "reward and punishment" is experienced as "blocks and flow." Because the researcher knows the desired end result, the problem is set up so that the machine experiences blocks and changes direction. This process of block and flow continues in a cyclical way as the machine learns proper selection of variables and spends its time looking for patterns only among those variables. Unlike in artificial intelligence, the machine cannot generate knowledge independently, but there is continual refinement of the methods of how the data is processed to examine it for patterns.

Understanding whether a problem is supervised, unsupervised, semi-supervised, or a reinforcement problem is the first step in initiating use of machine learning. It helps the researchers understand the problem well enough to make decisions about variable selection before analysis of the data. A full review of what happens after selecting the problem structure is beyond the scope of this chapter, as it brings us into a whole new discipline of computer science. We have included a brief introduction to problem structure here because understanding it helps researchers collaborate better with their colleagues in computer science. Collaboration among researchers in data science will be critical in advancing outcomes management from the current retrospective methods to methods that use automated predictive models for risk forecasting and proactive management of outcomes.

The next few sections of this chapter will review explorations into two common variables of interest. One is a review of predictive analytics studies on readmission for heart failure, and the other reviews studies using both predictive analytics and machine learning on reduction of falls. It is important to understand the current state of knowledge and theory as they relate to an outcome variable of interest since understanding what we already know about predictors will determine our problem structure for machine learning. If we have many predictors that are known, we can use a supervised or reinforcement problem. If there are only a few or no variables that have been determined to be predictors, then we would select a semi-supervised or unsupervised problem structure for our machine learning study.

Exploration of Two Variables of Interest: Early Readmission for Heart Failure and Falls

Before going into a review of how the Profile of Caring could be used in machine learning, we are going to look at two variables of interest discussed in Chapters 5 and 9 (falls and heart failure, respectively). Findings from the existing literature on how these two variables of interest have performed in predictive analytics studies could be used to set up a machine learning program to proactively manage these outcomes. As you read the literature review, notice the absence of references to the beliefs, behaviors, attitudes, or even the technical and/or relational abilities of the clinicians.

Readmission for Heart Failure

Of the two variables of interest reviewed for this chapter, readmission for heart failure was the only outcome for which there were systematic reviews. It also had the largest number of predictive models identified in the literature, which is significant, in part, because examination of most types of learning problems using machine learning is only possible after a predictive analytics model is set.[2] No machine learning studies could be found related to readmission for heart failure. The following review is included in this chapter to demonstrate the wealth of information available to researchers wishing to construct a machine learning study on readmission for heart failure.

Review of Predictive Analytics Studies Related to Heart Failure

Five reviews of predictive models regarding readmission for heart failure spanned the years 1948–2018 (Mahajan, Heidenreich, Abbott, Newton, & Ward, 2018). Mahajan et al. (2018) found a total of 329 papers in predictive analytics related to readmission to the hospital due to heart failure, and 304 of the studies examined a single predictor. The remaining 25 studies used multivariate models to evaluate what variables predicted readmission. Across the 25 models, there were 75 unique variables studied (Mahajan et al., 2018). Variables that were included most frequently in the 25 models included systolic blood pressure ($n = 9$), blood urea nitrogen ($n = 8$), history of heart failure ($n = 8$), serum creatinine ($n = 6$), hemoglobin ($n = 6$), number of times readmitted ($n = 6$), and age ($n = 6$).

Of the 75 predictor variables examined in the 25 models, none of the models included variables from current heart failure guidelines regarding the overall competence of caregivers or whether they were trained in heart failure. It is not clear whether these variables were reviewed as possible candidates for inclusion and failed to prove statistically significant predictors or if variables related to caregiver competency were never considered for inclusion in any structural models to test for the relationship they had with readmission for heart failure.

---

2 All types of learning problems except unsupervised problems require prior examination of predictors.

The final variables found to be statistically significant in the prediction of readmission for heart failure, in at least 1 of the 25 models as reported in the literature review by Mahajan et al. (2018), include variables organized into the domains of (a) clinical, (b) administrative, and (c) psychosocial. A full list of these variables organized by domain from these 25 models appears in Appendix D.

In hundreds of studies, models using predictive analytics to study readmission for heart failure were found to be generally useful and easy to apply, but models developed using machine learning were reported to be difficult to apply because of the large number of variables used in the models (Mahajan et al., 2018). Models that used secondary or randomized data were also difficult to use in machine learning studies (Mahajan et al., 2018). What is needed is automated extraction methods that produce information clinicians can immediately apply in operations (Mahajan et al., 2018). Retrospective data is typically plentiful, but unless it is being used in machine learning to understand trends over time to forecast risk, it is of limited value in predictive analytics.

What Successful Interventions for Reduction in Readmissions for Heart Failure Teach Us About What to Do Next

Review of the interventions shown to decrease readmission for heart failure supports the assertion of the authors of this chapter that integration of more data pertaining to the care process itself—such as competence, communication, and the quality of relationships among staff members—would make the models more precise and thus interpretable and applicable for just-in-time refinement of care. As you will see, the vast majority of the interventions found in the literature that demonstrated a reduction in readmissions for heart failure were those related to the communication skills of the clinician—a relational, interpersonal (rather than technical) factor. For this reason, the Profile of Caring is poised to become an essential variable in future studies on readmission for heart failure and other variables of interest.

Only one of the interventions addressed the technical skills of the clinicians involved: Use of home health nurses trained in heart failure demonstrated a decrease in readmissions from

29% to 16%, compared to nurses not trained in heart failure (Leavitt, Hain, Keller, & Newman, 2020).

All other interventions were interpersonal in nature. For example, a follow-up telephone call after discharge from hospitalization for heart failure reduced 30-day readmission by 54% (Reese et al., 2019). Mizukawa et al. (2019) tested an intervention to reduce readmissions due to heart failure that included a 30-min face-to-face meeting with a nurse once a month for 6 months followed by a telephone call every month for the following 6 months. In addition to the face-to-face and monthly telephone communication, Mizukawa et al. (2019) included remote monitoring of the patient's blood pressure, heart rate, and weight. If any changes were identified remotely, the nurse would call the patient to discuss these changes (Mizukawa et al. 2019). The intervention by Mizukawa et al. (2019) resulted in reducing readmissions for heart failure from 60% to 20%.

In a study by Agostinho et al. (2019), cardiologists followed up with heart failure patients after discharge at 1, 3, 6, and 12 months. The follow-up by the cardiologist included clinical assessment, lab work, electrocardiography, and assessment of adherence to and tolerance of therapy, titration of therapy as required, and whether the patient was performing self-care (Agostinho et al., 2019). The follow-up by cardiologists resulted in a reduction of readmissions for heart failure from 36% to 16% (Agostinho et al., 2019).

In a study by Hahn, Belisle, Nguyen, Alvarex, and Das (2019) an intervention by pharmacists was tested which also included follow-up with the heart failure patient after discharge. The pharmacists provided follow-up that related to an interdisciplinary plan of care versus merely educating the patient about medications, which is the standard of care for pharmacists (Hahn et al., 2019). This intervention led to a reduction in readmission for heart failure from 17.1% to 7.1% (Hahn et al., 2019).

Limitations of Past Work

The successful interventions were essentially all relational, but relational skills are largely ignored in the studies.

Given the high success rate of these interpersonal interventions to decrease readmission rates for heart failure patients, it is concerning that none of the predictive models designed to improve those outcomes tests staff behaviors, beliefs, or attitudes. It is perplexing that the successful interventions are essentially all relational, but relational skills are largely ignored in the studies.

Other concerns proposed by the authors pertain to the use of machine learning without theoretical guidance as to how to interpret the models. Theory in this regard can refer to knowledge of formal theoretical frameworks such as Watson's Theory of Transpersonal Caring (2008a) or Felgen's Theory of Clarity (Felgen & Nelson, 2016), or even a simple hunch proposed by the staff, or the belief that readmissions due to heart failure can occur because of the healthcare context. Finally, it is a limitation of the Hahn et al. (2019) study that the models are static. This limitation applies to any static model that is used in a dynamic environment. The models reviewed here tested results using a set of variables that are proposed to be predictors of readmission, but there is no acknowledgement of the dynamic, just-in-time information provided by the staff and/or system at various points in time. As predictors of readmission are identified and interventions to address the predictors are implemented, new predictors will surface. Such dynamics make static models inappropriate for consistent use over time as well as inappropriate for application to varied healthcare contexts.

Patient Falls

Examination of the predictive models for fall prevention reveals what is missing from the readmission literature for heart failure: an automated data collection method derived from a multivariate study of predictors. In the revolutionary work of Lee, Jin, Piao, and Lee (2016), inclusion of an automated data collection and reporting process resulted in a decision support system to prevent falls without nurses entering any additional data. The authors called this system the Automated Falls Risk Assessment System (Auto-FallRAS) (Lee et al., 2016).

Lee et al. (2016) used machine learning to collect, interpret, and apply real-time information from staff and system to make appropriate changes in the healthcare contexts they were analyzing. They used data from electronic medical records (EMRs) to compare patients who had experienced falls ($n = 868$) with a control group of randomly selected patients who had not experienced a fall ($n = 3,472$). There were 10 predictor variables in the final model. The final model was validated alongside the Morse Fall Scale (MFS), a commonly used fall risk assessment tool. Data for the MFS was collected concurrent to the EMR data collection tool. Sensitivity for the MFS was .68 while sensitivity for the Auto-FallRAS was an impressive .95. Specificity for the MFS was .60, and for the Auto-FallRAS, it was .50.[3]

The EMR was used to collect 4,211 variables for examination. Nurses were not required to enter data for the study since only EMR data was used. Data included 959 patient variables (e.g. admission variables, medical history, assessment); 19 environmental variables (e.g. ward, length of stay); 3,222 variables related to medical interventions (e.g., medications, treatments); and 11 nursing interventions (e.g., whether fall risk assessment was used). Of the 4,211 variables examined, 10 of them had statistically significant relationships with falls in a final multivariate regression analysis. The 4 steps used to drill down from 4,211 to 10 variables and interventions can be found in Appendix E.

Patients in the study were divided into three groups, according to risk. The patient's level of risk was evaluated each day at midnight. The results were put on the nurses' plan of care for the patient so the whole care-team could use the patient's risk level for decision making in the process of caring for patients. The high-risk group included patients who had at least one previous fall, patients who fell twice or more in last 6 months, and patients diagnosed with seizures. The moderate-risk group included patients under 7 years old, patients 65 years old or older, and patients on the psychiatric unit or pediatric ICU (Lee et al., 2016). The low-risk patients were completely paralyzed or immobile patients (Lee et al., 2016).

Limitations

Lee et al. (2016) identified as a limitation of this system of risk assessment that it did not update the level of severity of illness automatically; instead it is preferred that methods of continuous monitoring be provided to ensure more real-time data in acute care situations. Another limitation of this generally excellent study is the lack of assessment of the healthcare context itself and subsequent respecification of the model. It is the

---

3 Sensitivity refers to a measurement of how consistently the instrument correctly informs the clinician whether a fall will occur, while specificity refers to a measurement of how consistently an instrument predicts when a fall will not occur (Yip, Mordiffi, Wong, & Ang, 2016).

experience of the authors of this chapter that direct care providers often report idiosyncratic elements of the healthcare environment, which are not considered in static data collection tools, but which nonetheless impact outcomes such as falls. Ideally there would be a dynamic method to respecify the measurement model in real time using the input of the healthcare team.

The final limitation noted in the research in falls is the lack of assessing the disposition/profile of staff members, including constructs such as clinician competence, clarity of role, self-care, and job satisfaction. How are the energy and ability of the person providing the care impacting falls? What happens to incidence of falls when all staff members have clarity of role, when the primary nurse has clarity of role, or when nobody caring for the patient has clarity of role? We believe these things must be considered when measuring outcomes such as falls.

Proposal for a Machine Learning Problem

We propose constructing two machine learning-based analyses—one which combines the variables found in the Profile of Caring with the variables identified for readmission for heart failure and another that combines them with variables related to patient falls. It would seem that a supervised, semi-supervised, or reinforcement problem would likely be the problem structure of choice for a machine learning examination of falls reduction or readmission for heart failure since prior studies have yielded so much information about potential predictor variables for both outcomes. However, the best problem structure cannot be determined conclusively until the structural model is developed.

The Profile of Caring

As stated, the Profile of Caring is a trademarked set of variables used to describe nurses who are experiencing (a) clarity of role, (b) clarity of system, (c) caring for self, (d) the caring of the unit manager, and (e) job satisfaction.

A recent 8-country study of 2,046 nurses, which is described in Chapter 16 of this book, revealed 6 facets of the work environment that explained 82% of what comprises the latent variable of nurse job satisfaction. The six facets of nurse job satisfaction are satisfaction with (a) professional rewards, (b) communication with the unit manager/participative management, (c) patient care, (d) relationship with coworkers, (e) professional growth, and (f) autonomy. This eight-country study also found that nurse job satisfaction was predicted by whether the nurse practiced self-care, had a caring unit manager, and/or had clarity of role and clarity of system. In this same study, nurse job satisfaction was found to reduce nurse sick time, and the authors of the study propose that it will improve other patient outcomes as well.

Interestingly, the factors outlined in the Profile of Caring parallel the expected outcomes of many well-known frameworks of care delivery, such as Relationship-Based Care® (Creative Health Care Management, 2017; Koloroutis, 2004), Patient-Centered Care, Professional Practice Framework, and the Caring Behavior Assurance System©. These are examples of frameworks of care delivery that instruct care providers and leaders on what caring beliefs, behaviors, and ways of being are most desired for effective care delivery. Organisms, including people, evolve based on the environment they live in. If staff members and leaders are oriented and supported to behave and believe in a particular way, then the resultant observable behaviors which are presumed to advance the patient experience must be measured.

This section proposing a semi-supervised machine learning problem will include variables comprising the Profile of Caring as the supervised component, seeking to understand through machine learning, what we do not know about how to improve the patient experience.

Constructing the Study for Our Machine Learning Problem

For the purposes of this chapter, we will use what was learned from the falls literature to build a structural model for machine learning that includes the disposition of staff members, as measured by the Profile of Caring.

In Lee et al.'s examination of falls (2016), there were five administrative variables from environmental and medical domains (length of hospital stay, unit, number of beds in a room, number of daily tests, and medications used for neuro or circulatory issues) and five variables related to patient data (age, maximum heart rate, severe illness, activity, and hyponatremia). To these two groups of variables, we propose adding a third group which would include employees' report of clarity of role, clarity of system, care for self, the caring of the unit manager, and the six-facet construct of job satisfaction: (a) professional rewards, (b) communication with unit manager/participative management, (c) patient care, (d) relationship with coworkers, (e) professional growth, and (f) autonomy, all of which comprise the Profile of Caring. This last group of variables is designed to uncover what specific, identifiable clinician behaviors and beliefs lead to safer outcomes. A structural model to study falls using these three groups of variables is found in Figure 6.1.

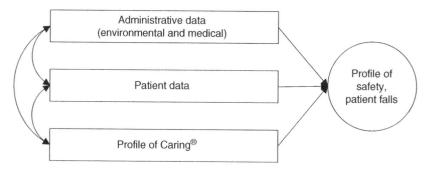

Figure 6.1 A structural model to study falls.

While the model in Figure 6.1 is much improved, it is still incomplete. Since we know that employees evolve over time, specifically during the implementation of a framework of care (Persky, Felgen, & Nelson, 2011), we also propose adding a fourth group of variables related to concepts taught in the framework of care and a fifth variable of time. The model in Figure 6.2 has been respecified to include these new variables.

At this point, the team would have become certain that an appropriate problem structure for our ML exploration would be semi-supervised, because we know some of the variables, but

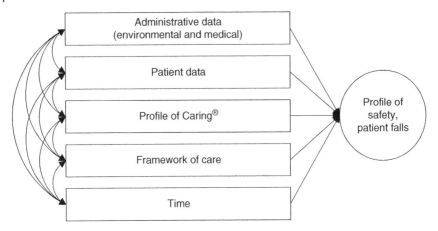

Figure 6.2   A structural model to study falls, respecified for implementing a framework of care.

we do not know the impact of the framework of care over time, and thus we would like the machine to teach us what it can about its impact.

Machine Learning Provides Data for the Real-Time Monitoring of Safety

As you can see in the model in Figures 6.1 and 6.2, we are solving for a variable we are calling the "Profile of Safety®" as it relates to patient falls. The Profile of Safety is a trademarked group of predictor variables, ranked from most influential to least influential, that comprise each safety profile the machine produces—essentially a list of factors that are most likely to predict falls given a particular group of patients in a particular place at a particular time. A Profile of Safety might tell us that, at this time on this unit, our most immediate fall risk is for people to whom psychotropic drugs are being administered, and it would go on to list the next most immediate risks in order of immediacy. In a few days, on the same unit, it is likely to tell us that we have a different set of risks. Because factors constantly change in the environment of care, the Profile of Safety is not static; it is always changing as predictors are mediated by interventions staff members implement and by changes in patient demographics. For example, if psychotropic drugs were the predictor, the staff might implement an educational program for all patients given psychotropic drugs and the side effect of dizziness, etc. Once the risk factor is mediated, it is no longer a risk factor, although this would only be true as long as the educational program was utilized. The risk factor of psychotropic drugs may resurface, but for the time of implementation, other factors will likely surface instead.

The benefit of using machine learning to manage risk is that you are not relying on outdated data—and in an assessment of risk for falls, yesterday's data is outdated data.

Using the Profile of Caring to Proactively Hire and Train Staff for Safety

While the Profile of Caring is just one set of variables contributing to the Profile of Safety, it has shown itself to be a very important one. The study that gave us the Profile of Caring, which is outlined in Chapter 16 of this book, determined that both nurse sick time and nurse turnover rates are improved when clinicians experience clarity of role, clarity of system, self-care, and the care of their unit managers. Outside of "care of the unit manager," these are all characteristics that can be identified in candidates for hire, and all four of these characteristics can be actively developed in healthcare environments.

As you recall in our review of the literature regarding readmission for heart failure in fewer than 30 days, all of the interventions that had proved effective were relational in nature. In other words, they were dependent on clinicians being knowledgeable, competent, and engaged. It could be said that what the Profile of Caring measures—(a) clarity of role, (b) clarity of system, (c) caring for self, (d) the caring of the unit manager, and (e) job satisfaction—is the degree to which clinicians understand their jobs and care enough about their work to do their very best. In short, clinicians who score high on the Profile of Caring are knowledgeable, competent, and engaged.

7

Forecasting Patient Experience: Enhanced Insight Beyond HCAHPS Scores

Mary Ann Hozak and John W. Nelson

The Hospital Consumer Assessment of Healthcare Providers and Systems (HCAHPS) survey is an assessment of the patient's experience of healthcare. It is mandated by the US federal government for all hospitals to use in measuring the patient experience. Each hospital's score is benchmarked with other healthcare organizations and is used for reimbursement and ranking of hospitals. It is important for hospitals to understand what *predicts* the patient experience in order to *improve* the patient experience. However, it is not easy to study predictors of the patient experience as articulated in the HCAHPS questions, since the US government does not allow organizations to use the HCAHPS questions without expressed approval by the government (CMS, 2018). Request for use of any item in the survey, any combination of items, or the whole survey must include what items you wish to use, how you intend to use them, information about the population to be studied, the purpose of the research, and the exact span of time over which the items will be used, as no portion of the HCAHPS survey is allowed to be used longitudinally (involving repeated observations of the same variables over short or long periods of time). This chapter reveals how an organization worked with the US government to study predictors of select HCAHPS questions to develop a proxy measure of the patient experience—measuring the same things the five HCAHPS items measured—which they could then use to study what predicts a good patient experience.

Using Predictive Analytics to Improve Healthcare Outcomes, First Edition.
Edited by John W. Nelson, Jayne Felgen, and Mary Ann Hozak.
© 2021 John Wiley & Sons, Inc. Published 2021 by John Wiley & Sons, Inc.

Methods to Measure the Patient Experience

The authors of this chapter contacted the HCAHPS project team at the Centers for Medicare and Medicaid Services (CMS) for permission to use the HCAHPS survey for examining the correlates of the patient experience. Several months later, we made a second call to CMS, letting them know we would be using only 5 of the 34 HCAHPS items. A week later, the study was approved by CMS, and we used the guidelines they provided to develop the protocol that was eventually used. It should be noted that a primary difference of this study, contrasted to how the HCAHPS items are typically used, is that this study was intended to inquire about the patient experience while the patient was still in the hospital, whereas the HCAHPS items are typically presented up to six months after the patient has been discharged from the hospital. A second difference was that the original HCAHPS survey used questions with a 4-point Likert scale for patients to respond, ranging from sometimes (1) to always (4). It was desired to have a more sensitive assessment of the patient experience, so a 7-point Likert scale was used, ranging from strongly disagree (1) to strongly agree (7).

The five items were selected based on what was of highest concern in the organization as it relates to nursing care as revealed by secondary analysis of data collected the previous year from 2,459 of the organization's patients.

The HCAHPS items about the patient's hospital experience were originally presented as questions but were changed to statements for the purposes of our study. The original HCAHPS items and their revised wording are as follows:

- During this hospital stay, did nurses listen carefully to you? Changed to: During this hospital stay, nurses listened carefully to me.
- During this hospital stay, how often did the hospital staff do everything they could to help you with your pain? Changed to: During this hospital stay, my caregivers did everything possible to help me cope with my pain.
- Before giving you any new medicine, how often did hospital staff tell you what the medicine was for? Changed to: Before giving me any new medication, my caregivers described possible side effects in a way I could understand.

- During this hospital stay, did doctors, nurses or other hospital staff talk with you about whether you would have the help you needed when you left the hospital? Changed to: During this hospitalization, my caregivers talked to me about how much help I might need when leaving the hospital.
- During this hospital stay, how often was your pain well controlled? Changed to: During this hospital stay, my pain was well controlled.

Along with the 5 HCAHPS items, the 10-item Caring Factor Survey© (CFS) was used to assess the 10 factors of caring—or "Caritas processes," proposed in Dr. Jean Watson's Theory of Transpersonal Caring (2008b). According to Watson, the 10 Caritas processes put into language the behaviors of caring that need to be measured if we want to know that caring is happening.

It was suspected that the 10 CFS items would relate strongly with the HCAHPS items and could possibly be used as a proxy measure for the five items of concern after the study period was complete since the HCAHPS questions could no longer be used beyond the study. Watson's 10 processes of caring are as follows:

1) Cultivating the practice of loving kindness and equanimity toward self and others. Loving kindness includes listening to, respecting, and identifying vulnerabilities in self and others.
2) Being authentically present: enabling, sustaining, and honoring faith and hope which is future-oriented and includes self-discovery.
3) Cultivating one's own spiritual practices and transpersonal self, going beyond ego-self.
4) Developing and sustaining a helping-trusting caring relationship.
5) Being present to, and supportive of, the expression of positive and negative feelings.
6) Creative use of self and all ways of knowing as part of the caring process; engaging in the artistry of Caritas (caring). At the core here is creative problem solving.
7) Engaging in genuine teaching-learning experience that attends to unity of being and subjective meaning: attempting to stay within others' frame.
8) Creating a healing environment at all levels.

9) Administering sacred acts of caring-healing by tending to basic needs.

10) Opening and attending to spiritual/mysterious and existential unknowns of life-death. This is belief in the impossible (miracles), even when others may assert doubt.[1]

Other variables that were measured included whether it was the patient or family member who responded, and in which hospital or on which unit the patient was treated. Data from families was measured to examine whether patient and family scores were similar enough that family scores could be used as a proxy for patients who were unable to respond, such as those patients who were too sick or too young to respond to a survey. Data specific to the hospital and unit was measured to help us understand where to concentrate efforts for operational refinement to improve the patient experience.

Fifteen items—the 5 HCAHPS items, rephrased to address the experience of people still in the hospital, and the 10 items in the Caring Factor Survey—were then presented to 232 patients just prior to discharge from the hospital. Once the data was collected, all 15 items would be examined together in a factor analysis. This would determine whether all 15 items loaded as a single construct or as 2 separate constructs. If they loaded as a single construct, that would mean each item had a statistically significant relationship to every other item among the 15, which would show that they are measuring the same thing.[2] If they loaded as a single construct, we were on our way to showing that if, after this study, we used only the Caring Factor Survey to measure care, we would be measuring the same things we would have measured with our five HCAHPS items.

To determine likely predictors of the patient experience, hierarchical regression was used, with the 5 HCAHPS items and the 10 CFS items as the dependent variables. Independent variables measured included race of patient, whether it was the patient or family member who responded, hospital, and unit. We wanted to see whether being of a particular race, being treated on a

---

1 Nelson, DiNapoli, Turkel, & Watson, 2011.
2 Oblique methods were used for extraction (direct oblimin) and rotation (principal axis factoring), Eigen values of 1.0 or more was used and Kaiser–Myer–Olkin (KMO) was used to assess for model fit.

specific unit, and/or having the survey taken by a family member influenced the findings.

Results of the First Factor Analysis

Responses to all 15 items were received from 214 of the 232 participants (92.2%). Results revealed that all 15 items loaded as a single construct and would not rotate, which confirmed that these 15 items are a single construct for this sample of respondents and not 2 separate constructs. The Kiser–Meyer–Olkin (KMO) measure of sample adequacy that is used to assess model fit was .95, indicating good fit. Table 7.1 provides the factor loadings of all 15 items. The higher the factor load value, the more important the factor is thought to be within the patient's experience. The highest loading items were "responded to me as a whole person" (an expression of holistic care) and "having a relationship with caregivers," which were both from the CFS. The third loading item, which was an HCAHPS measure, asked about feeling listened to by nurses. Factor loadings for all items were well above .40, indicating that each item was important in the measurement of the patient experience.

Implications of This Factor Analysis

The fact that an item that equates to "receiving holistic care" loaded first indicates that it is of highest importance among these 15 items. "Relationship" was a very close second, while the HACHPS items related to nurses listening, providing education, and controlling pain all loaded later. Conceptually, the order of these factor loadings suggests that holistic care serves as the foundation for excellent care as perceived by the patient and that a trusting relationship is the vehicle that drives delivery of what the HCAHPS items are measuring, which includes nurses listening, providing education, and controlling pain. Further investigation into these findings is warranted.

Examination of the correlation of the CFS total score and the HCAHPS total score revealed a score of .86 ($p = <.001$), indicating that the two tools did in fact measure the same thing. Chi square, another statistical method to assess "fit" of the CFS for use as a proxy, revealed that there was no statistical difference

Table 7.1 Factor loading of HCAHPS and CFS.

Survey and item wording	Factor loading
CFS6: My caregivers have responded to me as a whole person, helping to take care of all my needs and concerns	.924
CFS7: My caregivers have established a helping and trusting relationship with me during my time here	.904
HCAHPS1: During this hospital stay, nurses listened carefully to me	.892
CFS1: Every day I am here, I see that the care is provided with loving kindness	.890
CFS2: As a team, my caregivers are good at creative problem-solving to meet my individual needs and requests	.890
CFS4: When my caregivers teach me something new, they teach me in a way that I can understand	.869
CFS8: My healthcare team has created a healing environment that recognizes the connection between my body, mind, and spirit	.855
CFS3: The care providers honored my own faith, helped instill hope, and respected my belief system as part of my care	.796
CFS10: My caregivers are accepting and supportive of my beliefs regarding a higher power, which allows for the possibility of me and my family to heal	.792
CFS9: I feel like I can talk openly and honestly about what I am thinking, because those who are caring for me embrace my feelings, no matter what my feelings are	.790
HCAHPS3: During this hospital stay, my pain was well controlled	.785
HCAHPS4: During this hospital stay, my caregivers did everything possible to help me cope with my pain	.779
HCAHPS2: Before giving me any new medication, my caregivers described possible side effects in a way I could understand	.775
HCAHPS5: During this hospitalization, my caregivers talked to me about how much help I might need when leaving the hospital	.696
CFS5: My caregivers encouraged me to practice my own individual spiritual beliefs as part of my self-caring and healing	.664

between the responses from the patients ($n = 133$), family members ($n = 30$), or respondents who did not report whether they were the patient or family member ($n = 51$), suggesting that we could safely use the responses of family members as a proxy for patients who were too young or too sick to take the survey.

Predictors of Patient Experience

Using a regression equation, it was found that the unit on which someone was treated explained 19.1% ($p = <.001$) of the variance of the patient experience, using the combined CFS–HCAHPS score. It was noted in the details of the regressions (specifically, the coefficients), that Unit 1 and Unit 2, noted in Figure 7.1, had a statistically significant higher score than the reference unit 12 ($p = .003$ and .009, respectively).

Other variables that were examined included the patient's race and the specific facility in which they were cared for (there were two hospitals), but neither was shown to have a statistically significant relationship to the patient experience.

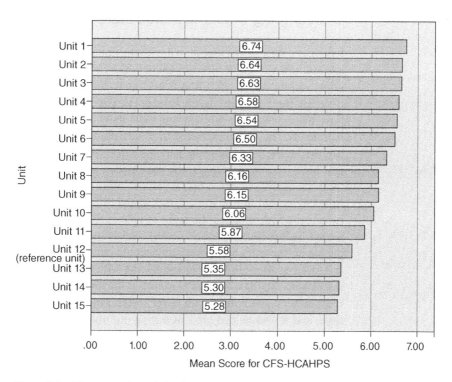

Figure 7.1   Mean score by unit for CFS-HCAHPS.

Discussion

While we were thrilled to learn that the CFS could be used as a proxy for five key HCAHPS items for as long a duration as we wanted to use them, the most significant finding to the organization was that the unit on which someone was treated predicted nearly 1/5 (19.1%) of the experience of the patient. We pondered what the staff profile of those units would be that creates a good experience, and we agreed that further study, next time incorporating the Profile of Caring®, a trademarked set of variables used to describe nurses who are experiencing (a) clarity of role, (b) clarity of system, (c) caring for self, (d) the caring of the unit manager, and (e) job satisfaction, was warranted, but that it is beyond the scope of this chapter.

The unit on which someone was treated predicted nearly 1/5 (19.1%) of the experience of the patient.

Transforming Data into Action Plans

Once the data analysis was shared with the unit advisory councils and the nursing leadership council, the performance and safety improvement (PSI) RNs, nurse managers, and clinical RNs took their unit-specific data and got to work to understand how these insights could inform their daily workflow and workloads. They examined what they were doing in care delivery, then reflected on how the patients perceived care delivery and how, in many instances, the patients felt it did not meet their needs, as indicated by responses to the five HCAHPs items measured in this study. Units scored differently in each category, which directed the team to look at those units who scored high (between six and seven), suggesting that perhaps people on those higher scoring units could counsel people on the lower scoring units about interacting with patients. Such discussions facilitated action planning at the unit level to promote better care for units with lower scores. This saved a lot of time, and it also helped prevent having multiple ways of providing care since this is an ISO accredited organization[3] and

---

3 International Organization for Standardization (ISO) is the independent, nonprofit, nongovernmental, international organization that sets the standards for operations in healthcare. Standard uniformity is facilitated by ISO tools that are developed collaboratively with all stakeholders. It is coordinated by a Central Secretariat in Geneva, Switzerland (*Source*: ISO.org).

standardization is considered a form of evidence-based practices across the system.

Summary

Predicting the patient experience using HCAHPS and a measurement instrument derived from Watson's Theory of Transpersonal Caring proved to be an invaluable process for this organization. It allowed staff members and leaders to prioritize where and how staff members needed to make the necessary changes for the greatest impact on the outcomes. It also highlighted which units were already doing exceptional work, allowing them to assume a leadership role with their peers. We have noticed that peer-to-peer projects tend to be more successful as they allow people to put aside doubts about whether interventions will work and experience more enthusiasm for the projects. It also helped that the organization celebrated the successes of the units with mean scores between six and seven, opening the door for more unit staff members to be celebrated as their work proved to sustain positive outcomes for patients and families.

The other element of this study that was helpful was educating nurses on what we learned. This data and associated use of the data are important for hospitals that are designated as or applying for Magnet® status. This hospital had received Magnet status four times at the time of this study and was planning to submit for redesignation. Predicting the patient experience using HCAHPS and a measurement instrument derived from Watson's Theory of Transpersonal Caring proved to be an invaluable process for this organization.

Predicting the patient experience using HCAHPS and a measurement instrument derived from Watson's Theory of Transpersonal Caring proved to be an invaluable process for this organization.

8

Analyzing a Hospital-Based Palliative Care Program to Reduce Length of Stay

Kate Aberger, Anna Trtchounian, Inge DiPasquale, and John W. Nelson

A collaborative project between medicine and nursing was undertaken to reduce hospital length of stay (LOS) and in-hospital death in the palliative care (PC) service line. This project took place 10 years after the organization's initial implementation of Relationship-Based Care® (RBC) (Creative Health Care Management, 2017; Koloroutis, 2004). Concepts from RBC, such as caring for self and others, shared governance, the I_2E_2 change model as described in Appendix B, and the importance of role clarity, were applied throughout the study to facilitate model development, data gathering, interpretation, and operational application of the findings.

Concepts from RBC were applied throughout the study to facilitate model development, data gathering, interpretation, and operational application of the findings.

Prior to the launch of the study, an examination of the literature helped us understand the existing research related to length of stay in palliative care. Cowen (2004) found that patients with advanced illness who were also offered palliative care had shorter lengths of stay, with an average of 5 days (range 1–48 days), when compared to patients with advanced illness who were not offered palliative care, who stayed an average of 11 days (range 3–114 days). Similarly, shorter lengths of stay for palliative care patients have been found in the medical intensive care unit (MICU) where these patients had an average of nine days in the MICU, versus the non-palliative care patients, who stayed in the MICU for 16 days on average (Norton et al., 2007). While some studies have shown that implementing a palliative care program reduces length of stay (Ciemins, Blum, Nunley, Lasher, & Newman, 2007;

Using Predictive Analytics to Improve Healthcare Outcomes, First Edition. Edited by John W. Nelson, Jayne Felgen, and Mary Ann Hozak. © 2021 John Wiley & Sons, Inc. Published 2021 by John Wiley & Sons, Inc.

May et al., 2017) and other studies have shown mixed results (Cassel, Kerr, Panilat, & Smith, 2010), there was ample evidence in the literature to suggest the possibility of reducing hospital length of stay through making changes in how palliative care was being used in this organization. One of the strategies that had proven successful according to the literature was to provide consultation for palliative care in the emergency department, close to when the patient arrived (Wu, Newman, Lasher, & Brody, 2013), or as early as possible after admission (May et al., 2017).

There was ample evidence in the literature to suggest the possibility of reducing hospital length of stay through making changes in how palliative care was being used.

The palliative care team had a strong hunch that shorter lengths of stay were happening in instances in which patients had palliative care consults fairly quickly after admission. As a strategy to reduce length of stay, the organization in this study sought to reduce two factors: (a) the time between admission and initial palliative care consultation, and (b) the time from initial palliative care consultation to the first patient/family planning meeting.

Building a Program for Palliative Care

The palliative care program in this 650-bed urban quaternary care center on the Northeastern Coast of the United States was started in 2008 by an advanced practice nurse (APN) who was certified in palliative care. She began by teaching the foundations of palliative care to the medical and nursing staff, using the End-of-Life Nursing Education Consortium (ELNEC) program. She also provided counseling to patients, families, and staff members in assisting with the development of end-of-life care plans. As the service line grew, they also hired a physician who specialized in palliative care.

In 2015 it became clear to this new team that a licensed clinical social worker (LCSW) would be needed to provide counseling and support to the palliative care patients and their families. By 2016 two more APNs were hired, along with an administrative assistant and Spanish interpreter, to assist with the evolving demands of the program. By 2017 there was an outpatient clinic program to meet the ongoing needs of the oncology patients enrolled in the palliative care program. As of 2018 the team consisted of one physician (clinical director), one RN (manager), two APNs, one LCSW, and one administrative assistant who also served as a Spanish interpreter.

The Context for Implementing a Program of Palliative Care

Those carrying out this study acknowledged that part of what made it successful was the work they had done while implementing Relationship-Based Care (RBC) and associated research (Hozak & Brennan, 2012). It was specifically what they learned about (a) care of self, team, and patient and family; (b) creating therapeutic systems; (c) role clarity; and (d) practice innovations and process improvement, that made the biggest difference. Several tactical interventions were used to integrate concepts of RBC into palliative care.

What made it successful was the work they had done while implementing Relationship-Based Care (RBC) and associated research.

Building a Model to Study Length of Stay in Palliative Care

Our existing palliative care audit tool was used as the initial model to study length of stay in the palliative care department. Data generated from the model would be used for analysis, discussion, interpretation, and application of findings to test the hypothesis, refine operations, and respecify the model used for the initial study. Data was collected from June to November 2016. The 14 predictor variables in Model 1, as they relate to the variable of interest, LOS, are noted in Figure 8.1.

Demographics of the Patient Population for Model 1

Data was collected on 286 patients, with 270 of the audits being complete for final analysis. Demographics were as follows:

- Respondents ranged from 31 to over 100 years of age, with the most prominent group being between ages 71 and 80 ($n = 81$).
- There were 16 distinct primary diagnoses with most reporting the diagnosis of cancer ($n = 102$).
- Within the diagnosis of cancer there were 17 different types of cancer reported, with lung cancer being the most commonly reported ($n = 26$).
- There were 23 different types or combinations of insurance reported, with the most common being Medicare ($n = 123$).

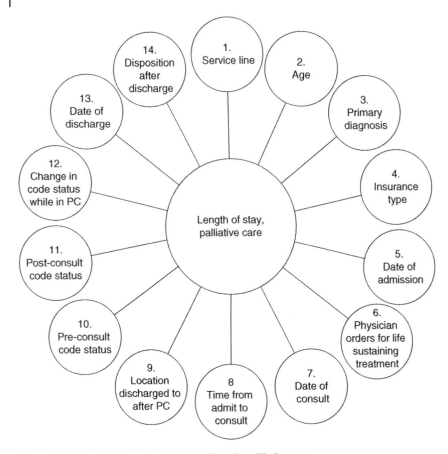

Figure 8.1 Model 1, to reduce length of stay in palliative care.

- When insurance was grouped by type, the largest remained Medicare only ($n = 123$), followed by Medicare and Medicaid ($n = 46$), Medicaid only ($n = 42$), commercial insurance ($n = 36$), with the remainder coded as "other" (e.g. charity care or combination of several types of insurance).
- Patients were from seven different clinical service lines, with the largest groups being from private ($n = 147$), hospitalist ($n = 74$), and medical service ($n = 29$).
- Most of the patients in palliative care were discharged without hospice care ($n = 102$).
- The most common code status prior to palliative care consultation was full code ($n = 170$), followed by do not resuscitate (DNR-B) ($n = 93$), which allows aggressive care, but not to the

point of cardiopulmonary resuscitation. Most of the patients remained full code from pre- to post-palliative care consultation ($n = 67$), with the second largest group changed from full code to DNR-B ($n = 63$).

- Only 47 of the 270 palliative care patients (17%) had the form completed for physician orders for life-sustaining treatments (POLST), which documents the patient's resuscitation wishes and goals of care.

Results from Model 1

For our study of Model 1, the variable of interest, length of stay, was calculated over a 5-month period, by subtracting the date of admission from the date of discharge. This calculation of days between admission and discharge revealed a range from 0 to 57 days, with a mean length of stay of 15 days, and a median of 29 days. Examination of the scatter plot of LOS over the 5-month study period revealed that the average length of stay declined from approximately 19 to 11 days during that time. This represented a 42% reduction in length of stay, and 2,032 fewer patient days (see Figure 8.2).

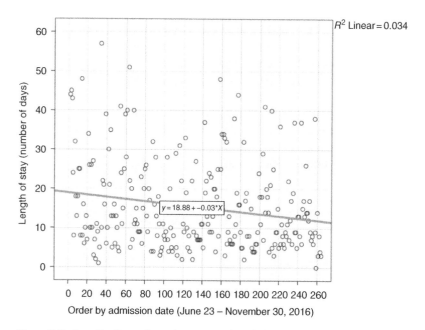

Figure 8.2   Length of stay throughout span of study for Model 1.

A 42% reduction in LOS resulted in $4.7–$5.6 million in total savings over the 5-month period.

At the time of this study, cost per patient day for hospital care in the state in which this organization resides was $2,349–$2,783 per day, which means that a 42% reduction in LOS resulted in $4.7–$5.6 million in total savings over the 5-month period.

In our study of Model 1, length of stay was examined in relationship to each of the predictor variables, using several regression equations. Among the 14 predictor variables noted in Model 1, three were found to have a statistically significant relationship with LOS. As suspected, "number of days from admission to time of palliative care consult" explained 34.1% ($p = <.001$) of the variance of LOS, and "service line" explained 11.5% of the variance ($p = <.001$). Patients in the medical service line had the longest LOS, while those in the hospitalist group had the shortest when compared to private care physicians who were not part of the hospital organization. The third variable to explain part of the variance of LOS was being discharged on hospice, explaining 4.7% of the variance of LOS ($p = .002$). Patients discharged to home on hospice had shorter LOS when compared to patients discharged while not on hospice ($p = .002$).

The three variables found to be most statistically significant in the individually run regression equations were then entered into a hierarchical regression with LOS as the variable of interest. The variables were entered in this order: "length of time from admission to consult," "service line," and "discharge on hospice." However, when we ran the final regression equation, it revealed that only two of these three were found to be statistically significant. "Length of time from admission to consult," on all units, explained 34.0% of the variance ($p = <.001$), while "service line" predicted 8.4% of the variance ($p = <.001$). Combined, these two variables explained 42.4% of the variance of LOS (see Figure 8.3).

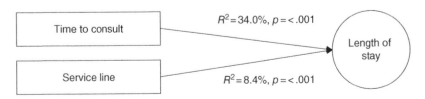

Figure 8.3   Results from Model 1: predictors of length of stay in palliative care.

Respecifying the Model

Model 1 was respecified in a second model for a follow-up study to include two additional variables deemed by the team as important to include in the model. These additional variables were identified during the discussion of results from the testing of Model 1. The two variables were "time from initial palliative care consult to planning meeting with the family" and "which physician referred the patient for enrollment in palliative care." (See Figure 8.4.) We had found that service line predicted LOS in Model 1, but we wondered if there were specific physicians within each

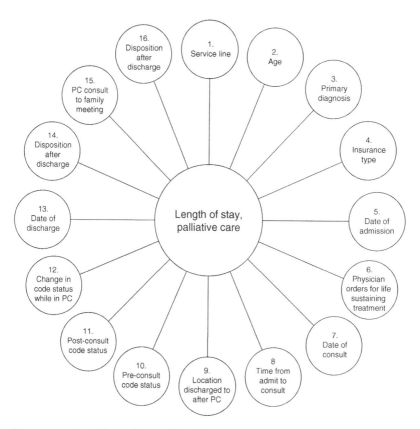

Figure 8.4 Model 2: predictors of length of stay in palliative care.

service line who were influencing LOS. Another 1,054 audits were conducted from January to December 2017 using Model 2. An examination of the frequencies of demographics determined that Model 2's demographics were similar to those in Model 1. Only age was noted to change slightly, with the age in Model 2 being slightly older, with the greatest frequency in the 81–90 age group ($n = 224$). Number of days from palliative care consult to patient/family planning meeting ranged from 0 to 29 days with a mean of 2.6 days. There were 127 physicians who referred patients to palliative care, with 7 of the physicians predominating the referrals within this group. The seven physicians responsible for most of the referrals were each coded for examination in the final regression equation. The intensivist group was the largest referring group ($n = 322$; 31%) and was thus divided into 3 types of intensivists, for a total of 10 individual or groups of referring physicians in the final analysis. Length of stay, as the variable of interest, ranged from 0 to 256 days. With these much higher numbers, when compared to Model 1, outliers (extreme values) were examined, and 39 patients with extreme values were removed as they were deemed risky to include as revealed by Cook's D, which provides an estimate of the influence of outlying data points when performing a regression analysis. Study of the outliers may reveal some valuable information, and a secondary analysis of the outliers would be beneficial.

Results from Model 2

Model 2 revealed that three variables predicted length of stay in palliative care. Similar to Model 1, "time from admit to initial palliative care consult" explained most of what predicted LOS, but to an even greater degree than it did in Model 1, predicting 64.8% of the variance of LOS ($p = <.001$). It was clear: The sooner palliative care staff members met with the patient, the shorter the length of stay. "Time from initial palliative care consult to patient/family planning meeting" explained 4.4% ($p = <.001$) of the variance, suggesting that the sooner palliative care staff met with the patient and family after the initial palliative care consult to create a plan of care, the shorter the LOS would be. Age predicted 3.1% of the variance ($p = <.001$), with higher age

It was clear: The sooner palliative care staff members met with the patient, the shorter the length of stay.

predicting shorter LOS. Patients in the 81–90 and 90-and-older age groups had shorter LOS than patients younger than 81 years of age ($p = <.001$). Combined, these three variables predicted 72.3% of LOS. We were somewhat surprised that no single physician was shown to be responsible for a longer or shorter length of stay considering the physicians were from varied clinical specialties. Final results from Model 2 are noted in Figure 8.5.

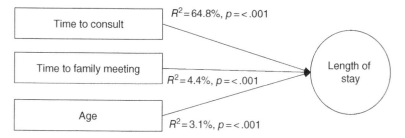

Figure 8.5 Results from Model 2: predictors of length of stay in palliative care with percent of explained variances.

Discussion

Operationally, the impact of the role of palliative care in reducing LOS in critical care was remarkable, not just in terms of utilizing resources, but also in some rather stunning cost savings. More importantly, it provided emotional support and comfort to patients and families in their most difficult times. The critical care environment can be overwhelming for patients and families due to the highly technical equipment, scientific vocabulary, and overabundance of alarms and monitors. Through their consulting with the patients, families, and physicians, this team offered patients and families comfort, additional information about their chronic or end-of-life diagnoses, and, most importantly, care-planning options to proceed with their healthcare choices. When the organization's leaders were informed of this study and the outcomes the palliative care professionals were able to produce, there was a palpable energy through the group to support them financially as a service line, even adding supportive positions for social workers, clerical staff, and additional APNs.

By studying the process with scientific rigor, we were able to identify and focus on the timing of the consult as a critical factor influencing length of stay, which prompted our team to increase education among the staff to encourage practitioners to initiate consults earlier. While cost effectiveness and efficient use of resources are operational goals, our first and most important goal is to treat patients with love, kindness, and human dignity, and to have them be an integral partner in the informed plan of care. The palliative care team held just such a vision, worked both hard and smart, and brought their vision of care to life.

Our first and most important goal is to treat patients with love, kindness, and human dignity.

One suggestion the authors have for readers of this chapter who are considering replicating this study is to measure the effectiveness of your framework of care. If you believe, for example, in the effectiveness of concepts taught in RBC, such as caring for self and others, clarity of role and system, and quality assessment using I_2E_2, then measuring the degree to which those concepts are visible in the culture would be important. Does time spent working in an RBC culture help increase the number of people with higher Profile of Caring® scores, as discussed in Chapter 6? Measurement of the framework of care would also answer whether the people who have received training in caring, clarity, and quality are the same people who contacted PC staff to come see the patient sooner, and it would answer how often it was the employees knowledgeable in these RBC concepts who helped set up the family meeting with the PC staff.

Additionally, measuring the healthcare team's knowledge of the framework of care using the Profile of Caring will reveal both the effectiveness of the new framework *and* the degree to which the people trained in the framework think and act in a more intentional, caring, and clear way for self and others. Chapter 18 of this book provides insight into how measurement of the operations of a framework of care can be accomplished.

9

Determining Profiles of Risk to Reduce Early Readmissions Due to Heart Failure

Mary Ann Hozak, Melissa D'Mello, and John W. Nelson

Predictive analytics have been used to study readmissions due to heart failure since 1948, and several reviews of existing models to predict readmission due to heart failure have identified 25 unique models that have been used with varied success (Mahajan, Heidenreich, Abbott, Newton, & Ward, 2018). Success in predicting readmission due to heart failure appears to be dependent on the inclusion of context-specific data that integrates administrative and clinical data, as well as the psychosocial data of the patient. (See Chapter 6 of this book for a comprehensive literature review on readmission for heart failure in fewer than 30 days.)

This chapter reviews the steps followed to develop a specified model for assessing readmission related to heart failure in a 650-bed acute care hospital in the Northeastern United States. The organization was struggling to reduce the number of readmissions in fewer than 30 days despite its use of a case study approach to reducing heart failure that had inspired a very high degree of engagement from the staff. While the team had succeeded in identifying a number of viable reasons for early readmission for heart failure, there was a feeling by the interdisciplinary heart failure team that they lacked a systematic way to identify and address the *most* common reasons for readmissions, and they were concerned that if they were not systematic in their approach, any improvements they experienced may just be coincidental, and therefore misleading. Thus, they hired an analyst who specializes in predictive analytics to develop a model specified for their context to reduce heart failure readmissions in their organization.

If they were not systematic in their approach, any improvements they experienced may just be coincidental, and therefore misleading.

Using Predictive Analytics to Improve Healthcare Outcomes, First Edition.
Edited by John W. Nelson, Jayne Felgen, and Mary Ann Hozak.
© 2021 John Wiley & Sons, Inc. Published 2021 by John Wiley & Sons, Inc.

The model in this study was based on a combination of the guidelines found in the literature for heart failure and the existing measurement instruments used by the organization to assess causes for readmissions for heart failure. The model was designed to allow the organization to use data to systematically evaluate all variables relating to readmissions in their organization, thereby allowing the organization to prioritize its actions to reduce readmissions. Predictive analytics were used here to identify the operational changes most likely to make the most significant positive impact on outcomes.

In the opinion of the authors, the most important part of this case study is that it shows how data contributed to meaningful conversations about what was contributing to readmission of heart failure patients in fewer than 30 days after discharge. As the people closest to the work examined their practice through the new lens provided by the data, three truths emerged: (a) every variable identified as a statistically significant predictor of readmission for heart failure in fewer than 30 days can and must be addressed with practice changes; (b) some variables that related to readmissions which fell short of statistical significance, but that connected to the existing hunches of staff, merit further study; and (c) variables that showed no statistically significant relationship with readmissions, even if contrary to the literature, do not merit practice changes at this time. Because the conversations were based in both the lived experience of the people closest to the work and information the data revealed about their own practice with their own patient population, there was tremendous buy-in for this process improvement project.

Step 1: Seek Established Guidelines in the Literature

The following three sets of guidelines helped us identify 98 variables reported to be important to the care of patients dealing with heart failure:

1) National Institute for Health and Care Excellence (NICE) Guidelines, 2014 and 2018
2) European Society of Cardiology (ESC), Acute and Chronic Heart Failure Guidelines, 2016
3) American College of Cardiology Foundation (ACCF)/ American Heart Association (AHA), 2013 and 2017 Guidelines for Management of Heart Failure

For a more thorough explanation of what was found in these guidelines, see Appendix F.

Step 2: Crosswalk Literature with Organization's Tool

The second step was to review the measurement instrument currently being used by the organization to collect data on their heart failure patients. Get With The Guidelines® (GWTG), which is the tool provided by the American Heart Association, was the foundation of this organization's tool. A few context-specific variables were added to the data collection process to customize the tool. There were a total of 121 variables in this tool, and 35 of them were also found in at least one of the guidelines listed in Step 1. This meant 86 of the tool's requests for data were unique to the data collection tool used by the organization. Unique variables included the names of specific cardiologists in the organization, distribution and/or explanation of specific educational material for patients, and whether the variable of interest—readmission in fewer than 30 days—was documented by the care provider.

It should be noted that during the course of this case study, using the model we developed, another literature review was conducted on all 25 predictive models identified by Mahajan et al. in 2018. Consequently, another 67 variables were identified by Mahajan et al. (2018) as relating to early readmissions for heart failure patients.

Step 3: Develop a Structural Model of the 184 Identified Variables

For those who do better seeing the variables and comparisons visually, a structural model was developed for the heart failure team. The model included 184 variables. There was some overlap between the guidelines and the organizational tool, as is noted in the crosswalk in Appendix G. The 184 variables in the structural model—98 from the literature and 86 additional variables from the organization's data collection tool—were color coded according to what was and was not currently being collected by the organization. Providing such a model served four purposes: (a) it enabled the team to see the complexity of studying an outcome like readmissions for heart failure; (b) it helped the team understand the focus of the data collected using the current tool; (c) it contrasted what was and was not being collected, which facilitated conversation about the data collection process; and (d) it facilitated conversations about how the model might be refined to more accurately measure the outcome of readmissions for heart failure in fewer than 30 days. A depiction of this comprehensive model—made to show the scale and complexity of our inquiry, but not used as a model in this study—is found in Appendix H.

The 184 variables were then compacted into 2 smaller models to help the heart failure team understand how the data was being categorized. We used the dimensions in the GWTG tool from the American Heart Association to make 11 categories. There are numerous ways to categorize the data, but we selected this method to align with the tool being used by the organization. The first model revealed the 11 dimensions of the GWTG tool, as well as 3 more areas that addressed contextual data related to (a) administration and hospital accreditation by Joint Commission, (b) arrival/admission data of the patient, and (c) patient demographics (see Figure 9.1).

A second model then broke the items down into more detail to help identify categories of items in each GWTG dimension to help us analyze which items were likely to be the most significant predictors of early readmission (see Figure 9.2).

In the model in Figure 9.2, the solid gray circles indicate all variables measured in the American Heart Association (AHA) heart failure tool (which was sometimes referred to as the Get

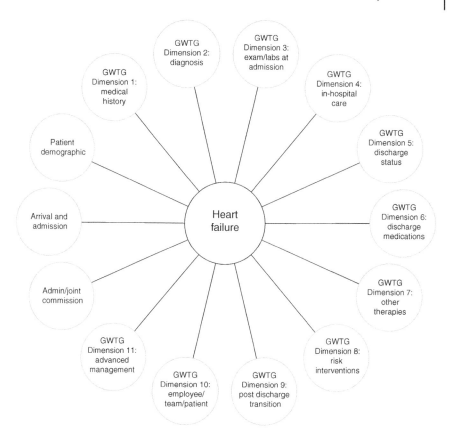

Figure 9.1   Model 1: organizational tool to examine heart failure readmissions.

With The Guidelines [GWTG] data collection tool), the one striped circle indicates a variable measured in part within the AHA heart failure tool which relates to the patient's past medical history, and the white circles indicate variables not currently measured in the AHA heart failure tool.

Step 4: Collect Data

We initially desired to automate the data extraction for this study with the help of an automated analysis program so predictors for early readmission could be understood in real time and even forecasted to manage practice proactively—before

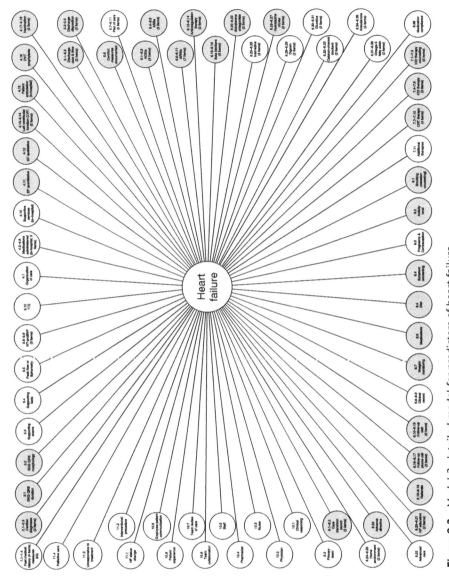

Figure 9.2 Model 2: detailed model for predictors of heart failure.

readmissions occurred. However, the queue of people waiting to have data extracted or interfaced was very long, and we were informed by the information technology people that automation would not be possible for quite some time. Because of the wait, we decided to collect data manually, and we realized that our best option was to use the organization's existing 121-item tool. Resources were tight, but the advanced practice nurse, trained in heart failure and designated to oversee practice related to heart failure patients, knew what data was needed; she used the model and collected data for 121 variables for a different heart failure patient every day she worked. In total, she collected data on 214 patients. Concurrent to collecting data for 121 variables on one heart failure patient each day, she also recorded admissions and readmissions of all heart failure patients. Collection of data on all admissions and readmissions helped us track the 214 patients as to whether they were readmitted and, if so, how many days after discharge the readmission occurred. Data related to the 121 variables was collected for 9 months for 214 patients. Data was collected for a full 12 months on all readmissions ($n = 430$) to enable the tracking of readmissions of the 214 patients with full data and to improve our understanding of the trends of readmission over the 12-month period of time.

Details of the Study

Regression analysis of all 121 variables in the data collection tool was used to study their relationship to the occurrence of readmission in fewer than 30 days. The study was limited to the fifth floor of the hospital, where patients with heart failure were admitted.

Variables for Which Data Was Collected

The following independent variables were examined in relationship to readmissions in fewer than 30 days.

1) Care provider
    a) Physician
    b) Cardiologist
2) Month of readmission
3) Frequency of readmissions that occurred in 30 days or more
4) Patient demographics
    a) Date of birth
    b) Age
    c) Race
    d) Gender
    e) Ethnicity
5) Payment type
    a) Payment source (listed first)
    b) Payment source (listed second)
    c) Payment service
6) Medical history
    a) Specific diagnoses
        i) Medical history: Afib
        ii) Medical history: Chronic obstructive pulmonary disease (COPD) or Asthma
        iii) Medical history: Heart failure
        iv) Medical history: Hypertension
        v) Medical history: Hyperlipidemia
        vi) Medical history: Coronary artery disease (CAD)
        vii) Medical history: Diabetes (non-insulin dependent)
        viii) Medical history: Anemia
    b) Known history of heart failure prior to this admission
    c) Medical history: Number of diagnoses listed
    d) Number of diagnoses of seven most commonly reported
7) Smoking
    a) History of smoking in the last 12 months
    b) History of smoking
8) Vital signs
    a) Heart rate
    b) Blood pressure: Systolic
    c) Blood pressure: Diastolic
9) Serum Creatinine
10) EKG
    a) EKG QRS in ms
    b) EKG Morphology (e.g. LBBB, RBBB, etc.)

11) Heart as a pump
   a) Ejection Fraction (EF as a percent)
   b) LVSD
12) Follow-up
   a) Follow-up visit scheduled
   b) Follow-up phone call scheduled
   c) Advanced care plan discussed
13) ICD 9 code
   a) ICD 9 code 42823
   b) ICD 9 code 42833
   c) ICD 9 code 42843
14) ICD 10 code
   a) ICD 10 code I5023
   b) ICD 10 code I5033
   c) ICD 10 code I5043
15) Medications
   a) Hydralazine nitrate prescribed
   b) Aldosterone antagonist prescribed
   c) Beta blocker prescribed
   d) Anticoagulation medication prescribed
   e) ARNI prescribed
   f) ARB prescribed
   g) ACE inhibitor prescribed
16) Patient's discharge disposition on the day of discharge
   a) Discharged to home
   b) Discharged to acute care
   c) Discharged to other facility
   d) Expired
   e) Hospice at home
17) Vaccine
   a) Pneumococcal vaccine received
   b) Influenza vaccine received

Zeroing in on Frequency of Readmission

Over the 12-month evaluation period, there were 430 patients readmitted for heart failure. An ANOVA procedure[1] was used to examine whether any month over the year-long study had

---

[1] An ANOVA procedure is an analysis used to determine whether there are statistically significant differences between groups.

a statistically significant difference from the other months in the number of days between discharge and readmission. No month was found to have a statistically significant difference in number of days ($p = .596$) between discharge and readmission. The average number of days and standard deviation remained unchanged from a statistical point of view. Number of readmissions and days between discharge and readmission are noted in Table 9.1.

A bar graph was also generated to depict both readmissions 30 days or more after discharge and readmission in fewer than 30 days after discharge. Results revealed that readmissions in fewer than 30 days had a greater frequency than those occurring after 30 or more days (see Figure 9.3).

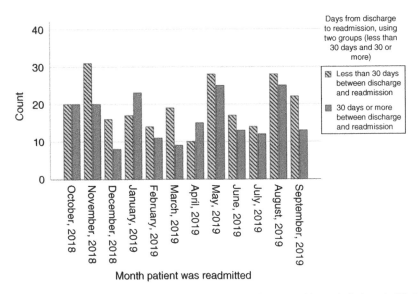

Figure 9.3 Comparing readmissions in fewer than 30 days with readmissions in 30 days or more.

This unfortunate finding was true for 10 of the 12 months examined. Cross-tabulation analysis was used to see whether there was any difference in overall readmission rates between these two groups when comparing months. No statistically significant difference was found between these groups ($p = .866$), indicating that people readmitted in fewer than 30 days and

Table 9.1 Number of readmissions and number of days from discharge to readmission.

Month	Number of readmissions	Mean number of days discharge to readmit	Std. deviation	Min days discharge to readmit	Max days discharge to readmit
October	40	24.55	18.83	1	62
November	51	33.37	26.568	2	91
December	24	30.29	24.736	1	73
January	40	33.48	22.929	1	73
February	25	27.96	23.851	1	84
March	28	24.61	23.186	1	75
April	25	39.00	22.411	3	81
May	53	30.60	22.656	1	84
June	30	29.80	23.330	4	79
July	26	34.08	24.280	1	89
August	53	32.38	23.647	2	87
September	35	29.60	25.789	2	90
Total	430	30.85	23.577	1	91

people readmitted in more than 30 days had roughly the same number of readmissions every month.[2]

In this study of the 214 patients on whom full data was collected, there were 56 patients who were readmitted in fewer than 30 days at least once. Most were readmitted in fewer than 30 days just once ($n = 44$), but there were 7 patients who were readmitted in fewer than 30 days twice, and 5 patients who were readmitted in fewer than 30 days three times during the study period (see Table 9.2).

Table 9.2 Frequency of readmission in fewer than 30 days.

Frequency readmission in fewer than 30 days	Number of patients	Percent
0	158	73.8
1	44	20.6
2	7	3.3
3	5	2.3
Total	214	100.0

Among the 214 patients for whom complete data was collected, there were 356 readmissions throughout the 12 months of the study, with an average rate of 24.6 readmissions per month. The average individual patient was readmitted 1.7 times with a standard deviation of 1.08 and range of 1–9.

Limitations of the Study

Data used to analyze what predicts frequency of admissions is limited in that only one set of data was collected for each patient. An inability to collect the data for each patient upon every admission and every readmission is a limitation of this study— and likely a significant limitation, in that it allowed only partial study of the patients' change in treatment and physical status.

---

2 Patients who are readmitted in fewer than 30 days after hospital discharge, specifically those who are on Medicare, are not reimbursed for by Medicare. This is a big impetus for organizations to decrease their readmissions in fewer than 30 days.

Results: Predictors of Readmission in Fewer Than 30 Days

This section includes (a) a brief summary of variables that proved insignificant after analysis, (b) a summary of variables that yielded inconclusive findings that did not warrant further study, (c) a summary of variables that yielded inconclusive findings that did warrant further study, and (d) an amplified description of variables that proved significant as predictors of heart failure patients being readmitted in fewer than 30 days.

Summary of Variables That Proved Insignificant After Analysis

The following variables showed no significant influence on whether heart failure patients were readmitted in fewer than 30 days after discharge: the patient's age; gender; history of smoking; vital signs (heart rate and blood pressure); serum creatinine; EKG; payment type; whether a follow-up visit was scheduled; ICD codes 9 and 10; whether hydralazine nitrate, aldosterone antagonist, beta blockers, anticoagulation medication, ARNI, ARB, or ACE inhibitors were prescribed at discharge; and whether the pneumococcal or influenza vaccines were administered.

Details about these findings can be found in Appendix I.

Summary of Inconclusive Findings

For a variety of reasons, the following variables produced inconclusive findings: the patient's race, ethnicity, smoking cessation, and activity level. In regard to the patient's race and ethnicity, insufficient data was collected; for example, the ethnicity of 161 out of 214 patients was labeled "No/unable to determine." Due to the small sample size, the effects of smoking cessation and activity level were not examined.

More about these findings can be found in Appendix J.

Description of Inconclusive Findings Warranting Further Study

While the variables of discharge disposition and who the patient's cardiologist was technically led to inconclusive findings, we learned some things about both of them that merit further study with a larger sample size and more complete data.

Discharge Disposition

Most of the 214 patients were discharged to home ($n = 160$). The second most common discharge disposition was "other health care facility" ($n = 34$). All categories and associated frequencies are noted in Table 9.3.

Table 9.3 Disposition of patient upon discharge.

Discharged to:	Frequency	Percent
Home	160	74.8
Hospice—home	5	2.3
Acute care facility	5	2.3
Other healthcare facility	34	15.9
Expired	7	3.3
Left against medical advice/AMA	2	.9
Total	213	99.5
Data missing	1	.5
Total	214	100.0

Being discharged to home was the reference, so all other categories were compared to home. Examination of the coefficients revealed that those who were discharged to an acute care facility ($n = 5$) were most likely to be readmitted ($t = 2.03$, $p = .044$). Examination of the five people discharged to acute care facilities revealed that three of the five people each had two or more readmissions and two of the five people each had two admissions in fewer than 30 days. Using linear regression, this frequency of five is a limitation. When we used Parato Mathematics to study outliers (and we considered these five people outliers) we learned that the variable of patients being discharged to acute care settings warrants further study, as patients in this group may be at higher risk for readmission in fewer than 30 days.

Cardiologist

The variable of who the patient's cardiologist was, while proving inconclusive in this study as a predictor, yielded some information we think is worth further study. When we looked at who the cardiologist was on each readmission ($n = 430$), a regression equation was used to study whether the cardiologist (those with over 20 readmissions) had any relationship to number of days from discharge to readmission. The reference group was "all other" cardiologists with fewer than 20 readmissions. There was no relationship found ($p = .732$) between cardiologist and rate of readmission (see Table 9.4).

Table 9.4 Readmission by cardiologist from 12-month readmission data set.

Cardiologist	Frequency of admissions in fewer than 30 days	Percent of total patients
Cardiologist 1	26	6.0
Cardiologist 2	61	14.2
Cardiologist 3	25	5.8
Cardiologist 4	31	7.2
Cardiologist 5	44	10.2
Cardiologist 6	27	6.3
Cardiologist 7	21	4.9
Cardiologist 8	23	5.3
Cardiologist 9	24	5.6
All others	148	34.6
Total	430	100.0

However, it should be noted that examination of the coefficients and t-values greater than 1.0 revealed that there were two cardiologists whose lower readmission rates were close to being statistically significant; this included cardiologist 6 ($t = -1.72$, $p = .087$) and cardiologist 3, ($t = -1.403$, $p = .162$). In contrast, cardiologist 2 had higher readmissions ($t = 1.16$, $p = .247$). This higher readmissions rate did not reach statistical significance, but considering the small sample and review of readmission numbers in Table 9.4, the likelihood of this reaching statistical significance in a larger sample is high. This implies that further study of cardiologists' rates of readmission is warranted.

Variables That Proved Significant Predictors of Readmission for Heart Failure in Fewer Than 30 Days

The following five variables were shown to have significant influence on whether heart failure patients were readmitted in fewer than 30 days after discharge: (a) number and type of concurrent diagnoses/comorbidities, (b) left ventricular systolic dysfunction (LVSD), (c) an ejection fraction of 40% or less, (d) total number of readmissions, and (e) whether a follow-up phone call was scheduled. As these were the most significant predictors of readmission in fewer than 30 days, we will look at each variable in more detail.

Concurrent Diagnoses/Comorbidities

There were 26 different options for concurrent diagnoses from the GWTG document, with 7 of the diagnoses being reported by enough people in the sample to examine them in a regression equation. The seven diagnoses most commonly reported are noted in Table 9.5.

Table 9.5 Frequency of specific diagnoses concurrent with heart failure.

Diagnosis	Frequency	Percentage of patients reporting
Atrial fibrillation	118	55.1
COPD	103	48.1
History of heart failure	174	81.3
Hypertension	189	88.3
Hyperlipidemia	148	69.2
CAD	100	46.7
Diabetes (non-insulin dependent)	90	42.1

Examination of each concurrent diagnosis in a correlation equation revealed that two of seven concurrent diagnoses, as noted in Table 9.5, had a statistically significant relationship to readmission frequency in fewer than 30 days, including COPD ($r = .207$, $p = .002$) and CAD ($r = .166$, $p = .016$), which were found to be the most significant predictors.

Frequency of total concurrent diagnoses among the seven was also examined, with patients having anywhere from one to seven diagnoses (see Table 9.6). Regression analysis of frequency of concurrent diagnoses revealed that concurrent diagnoses predicted 4.3% ($p = .002$) of readmission for heart failure in fewer than 30 days.

To be clear, there are two findings relative to concurrent diagnoses that were found to be statistically significant predictors of readmission for heart failure in fewer than

Table 9.6   Frequency of concurrent diagnosis.

Number of concurrent diagnoses (among seven most common)	Frequency	Percentage of patients reporting
1.00	15	7.0
2.00	25	11.7
3.00	24	11.2
4.00	41	19.2
5.00	50	23.4
6.00	37	17.3
7.00	21	9.8
Subtotal	213	99.5
Missing	1	.5
Total	214	100.0

30 days: (a) having *any* comorbidities (but especially COPD and CAD) increased a patient's likelihood of being readmitted in fewer than 30 days, and (b) the *more* comorbidities a person had, the more likely they were to be readmitted in fewer than 30 days. This means that interdisciplinary care teams caring for patients with heart failure should pay particular attention to patients with COPD and CAD, as well as to patients with multiple comorbidities.

Left Ventricular Systolic Dysfunction (LVSD)
There were 213 of 214 patients for whom information on LVSD was available. Most reported "yes" to the question about LVSD ($n = 151$), while 62 reported "no." Examination of this independent variable in a regression analysis of readmission in fewer than 30 days revealed LVSD to have a statistically significant relationship with readmission, predicting 3.4% of the variance of readmission in fewer than 30 days ($p = .007$).

Ejection Fraction
All 214 patients reported ejection fraction (EF) with a range of 15–70% and a mean percentage of 42% (standard deviation 16.13).

Correlation of EF with readmission in fewer than 30 days revealed a statistically significant negative relationship ($r = -142, p = .038$). This negative relationship means that the higher the ejection fraction, the lower the frequency for readmission in fewer than 30 days. Looking deeper at this finding using regression analysis, it was found that EF predicted 2.0% of the variance of readmission in fewer than 30 days ($p = .038$).

For both LVDS and low EF, the following actions can be implemented:

1) Lifestyle Changes
   a) Control of risk factors (i.e., healthy diet, exercise, no smoking, healthy weight)
   b) Patient education and participation in disease management program (i.e., follow-up appointments, phone calls).
2) Pharmacological Intervention (Medications)
   a) Diuretics for symptom relief
   b) ACE inhibitors/ARB/ARNI to widen blood vessels
   c) Beta-blockers to help decrease demands on the heart
   d) Nitrates to work to relax the blood vessels
   e) Inotropes to help heart pump harder
3) Surgical Interventions
   a) Implantable cardioverter defibrillator (ICD), cardiac resynchronization therapy (CRT)
   b) Left ventricular assist device (LVAD)
   c) Surgery is not frequently used to treat heart failure but to treat underlying cause
      i) Transplant
      ii) Percutaneous coronary intervention (PCI or angioplasty)
      iii) Coronary artery bypass
      iv) Valve replacement

Total Number of Readmissions

During the study, there were 44 patients who had at least one readmission after 30 or more days after discharge. We wanted to know whether those who were readmitted 30 days or *more* after discharge were also the same patients who had the readmissions in fewer than 30 days. A regression equation revealed that having been readmitted after 30 or more days predicted 2.8% of subsequent readmissions in fewer than 30 days ($p = .014$). This means that those who are being admitted in fewer than 30 days after discharge are also the patients who are more likely to be admitted 30 or *more* days after discharge. They were just plain more likely to be readmitted.

However, it should be noted that "readmission after 30 days or more" was entered into a regression equation as a second step, after a regression equation was run on concurrent diagnoses. You will recall that concurrent diagnosis explained 4.3% of the variance, meaning that concurrent diagnoses alone were statistically significant, and readmission after 30 days or more did not prove statistically significant when these two variables were studied separately. When they were studied together, however, the explained variance for concurrent diagnosis increased from 4.3% to 4.8%, indicating that if patients are coming in at 45 or 90 days, it means they are at a higher risk of coming in in fewer than 30 days, too, especially if they have concurrent diagnoses.

The finding that diagnoses other than those directly related to heart failure appear to be predictors of readmission in fewer than 30 days suggests that a team that includes clinicians beyond just cardiologists may help reduce early readmissions. These teams could create and operationalize a "complex patient" care plan to address each heart failure patient's multiple diagnoses. Within the structure of these interdisciplinary teams, a process could be created through which all specialists who saw each patient in previous admissions would have access to the patient's full care plan. CNSs who know each patient would also be invaluable members of these teams. Implementation of Primary Nursing (Wessel & Manthey, 2015), medical home models, and/or a case management structure can also be established to address the coordination and continuity of all medical and nursing interventions.

A team that includes clinicians beyond just cardiologists may help reduce early readmissions for heart failure.

Follow-Up Phone Call Scheduled

This variable was found to be the most significant predictor (or more accurately, preventer) of readmission in fewer than 30 days after discharge. Of the 214 patients, 60 (28%) had a follow-up phone call scheduled, 146 (68%) did not, and 8 did not report whether a phone call was scheduled. Examination of this variable using regression analysis revealed that the scheduling of a follow-up phone call had a statistically significant negative relationship with readmissions in fewer than 30 days, meaning that those who had a follow-up phone call scheduled were less likely to be readmitted within 30 days post discharge. This explained 5.9% of why patients were readmitted in fewer than 30 days post discharge ($p = <.001$).

This means that in a sample size of 1,000 patients, if we changed only the variable of adding a follow-up phone call, we could expect 59 fewer patients to be readmitted in fewer than 30 days, with a cost savings for the organization of $9,051 per readmission for heart failure (Mayr et al., 2017). In this organization, which serves roughly 400 heart failure patients per year, instituting the very low-cost intervention of adding a follow-up phone call would result in a savings of $213,603.

It should be noted that we specifically measured whether a follow-up call was scheduled, not whether the follow-up call occurred. It is a reasonable hypothesis that if we were able to look at readmission outcomes in instances where follow-up calls were completed versus not completed, the explained variance would be even higher. In future studies, it is recommended that analysts collect data on completion rates of follow-up calls in order to better measure the impact of follow-up calls on outcomes related to readmission in fewer than 30 days for heart failure patients.

Next Steps

These findings suggest that several changes in practice would likely cause a significant drop in readmissions in fewer than 30 days and thereby remove a significant financial burden from the organization. In discussing the results of the heart failure study and the five variables which had significant influence on readmission in fewer than 30 days after discharge, the cardiology team concentrated on the development of a robust heart failure program focused on continuum of care, to help patients from initial diagnosis of heart failure through lifestyle management with consideration for any concurrent diagnoses and comorbidities. The question we asked ourselves was, "Why are we concentrating on just preventing a 'readmission' when perhaps we should concentrate on preventing that first admission?"

As we developed a charter for this new program we formulated the following goal statement:

Develop and implement a system-wide structure and set of interventions to facilitate communication between clinicians in different care environments with consistent approaches to patient treatment, education, and emotional support to improve patient outcomes though healthy lifestyle management.

Our challenge was appropriating and leveraging the needed resources across the continuum for both outpatients and inpatients to achieve the goals through the following objectives:

- Reduce the percentage of heart failure patients readmitted to the hospital within 30 days of discharge.
- Identify potential gaps in transitional care for patients with heart failure and any co-morbidities that contribute to potentially preventable readmissions and enhance their quality improvement efforts.
- Provide education and support on advance care planning, emphasizing its importance with patients and caregivers to live a healthier lifestyle.
- Enhance patient self-management.
- Standardize heart failure treatment protocols across the continuum.

The program is designed to be driven by advanced practice nurses (APNs) assuming the primary caregiver role. Under the auspices of the Medical Director for Heart Failure and the chairperson for the Department of Cardiology, these APNs assess, evaluate, and coordinate the patients' care in collaboration with the patients' attending physicians and specialists. A nurse navigator also routinely keeps in touch by phone, email, and text with patients and families to see if they have questions or needs while at home. A data abstractor maintains program metrics and assists in collecting the data for the American Heart Association's Get With The Guideline program, working with the information technology department to develop daily push reports from the electronic medical record.

Patients are referred into the program at the time of their initial diagnoses of heart failure. Education, support, and appropriate medication management are provided to these families to engage them in healthy lifestyles and self-management. Cardiac rehabilitation is offered to assist with increasing stamina and strength, diet management, and pulmonary rehabilitation if comorbidities such as COPD exist. This program encompasses caring for the whole person and family without isolating a singular point in the disease process as causing an admission or a readmission.

10

Measuring What Matters in a Multi-Institutional Healthcare System

Kay Takes, Patricia Thomas, Gay Landstrom, and John W. Nelson

This chapter reveals how a multi-institutional healthcare system, comprising 94 hospitals in 22 states, including 109 continuing care and home care locations, studied the construct of caring while considering hospital individuality, including the effect of using Relationship-Based Care® (Creative Health Care Management, 2017; Koloroutis, 2004) as their care model. The study examined how the addition of variables for "clarity of self, role, and system" to an initial model testing caring for self and job satisfaction helped explain the operations of caring for patients.

A significant challenge in conducting a multi-system, multi-hospital organizational study was that multiple theories of caring were in use in the various organizations. Theories of caring used across the organization included Watson's Theory of Transpersonal Caring (Watson, 1979, 1985, 2008a), Duffy's Quality Caring Nursing Model (Edmundson, 2012), Swanson's Theory of Caring Behaviors (1999), and Caroline Coates' theory (1997) which combines Watson's Theory of Transpersonal Caring with Bandura's (1977) concept of self-efficacy. Despite the variations in the concept of caring for each theory, all of these theories have the same general purpose, and it is suggested that if clinicians behave according to any of these theories of caring, caring will be perceived by the recipients of those behaviors.

The researchers in the study accessed six tools to assess caring—three to assess caring as perceived by patients, and three to assess caring as perceived by caregivers—as well as three more tools which measure self-care; nurse job satisfaction; and clarity of self, role, and system. An effort was made to use tools that reflected all of the models of caring used in this large healthcare system. A comprehensive and eye-opening description of the tools used in this study can be found in Appendix K. In the process of refining and validating these tools, the data analyst and research team discovered some compelling information about the differences in perceptions between staff members and patients.

Testing a Model of Caring

Model 1, tested in the first year of this study, had two parts. The first part of Model 1 examined the relationship between self-care and job satisfaction. Job satisfaction was defined in this study as "the feeling derived from perceiving that the social and technical aspects of the work environment are sufficient to perform the job" (Nelson, 2013, p. 23). It was hypothesized for this study that staff members who rated themselves high in self-care would also have high job satisfaction.

It was hypothesized for this study that staff members who rated themselves high in self-care would also have high job satisfaction.

The second part of Model 1 examined whether those who reported higher job satisfaction also enacted more caring behaviors with their patients. The presence or absence of caring behaviors with patients was to be examined from the perspectives of both the staff members who cared for the patients and from the perspectives of the patients themselves. In the second part of Model 1, the research team wanted to understand whether there was a statistically significant relationship between what staff members and patients reported. The initial hypothesis was that staff members and patients would report similar perspectives on the quality of the care provided and received. Figure 10.1 shows Model 1, which was tested in the first year of this study.

The research team wanted to understand whether there was a statistically significant relationship between what staff members and patients reported.

Methods

A cross-sectional study was conducted in nursing services of 14 hospitals in the healthcare system which grew from 56 to 84 hospitals over the time of the study. At the time of the writing of this chapter, the system had grown to 94 hospitals. Ten hospitals from 4 states were involved in the baseline assessment which included 2,831 nurses on 96 units. There were 734 patients invited to participate.

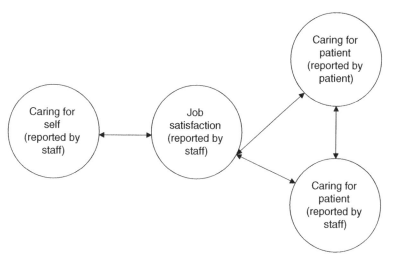

Figure 10.1   Model 1, to test relationships of self-care, jobs satisfaction, and quality of patient care.

It was not possible to examine directly whether there was a correlation between nurse job satisfaction and the patient's perception of caring, because we did not have consistent dyads of nurses and patients. However, we were able to measure the patient's perception of caring at the unit level and job satisfaction of nurses at the unit level, so we could assess whether the patients' perceptions of the caring behaviors were high on a specific unit, then look at whether job satisfaction was also high on that same unit. If patients were reporting their perceptions of caring behaviors and staff members were reporting their perceptions of job satisfaction on the same units, we could examine whether the patient's report of caring increased when staff job satisfaction increased.

Thirty-three of the 96 units collected data from patients. Six-hundred-twenty of the 734 surveys were completed, for a completion rate of 85.5%. Four-hundred-forty-four staff members responded on these 33 units. Relationships between all variables in Model 1, using data from these 33 units, were analyzed using Pearson's correlation. An alpha of .10 was used to adjust for the smaller sample size of units. Measurement instruments for all of the theories of caring, clarity, and job satisfaction are described in Appendix K.

Results from the Study of Model 1

Testing on the first part of Model 1, which looked at the relationship between self-care and job satisfaction revealed that there is, in fact, a positive relationship between self-care and job satisfaction ($r = .382, p = .028$).

There is, in fact, a positive relationship between self-care and job satisfaction.

Testing on the second part of Model 1, which looked at whether clinicians who are satisfied in their jobs enacted caring behaviors, however, revealed that there was no relationship between job satisfaction and the enactment of caring behaviors as reported by the staff ($r = .58, p = .750$) and no relationship between job satisfaction and enactment of caring behaviors as reported by the patient ($r = -101, p = .574$). There was, however, one statistically significant relationship found in the second part of the model: Staff members and patients reported "caring for the patient" similarly ($r = .486, p = .004$), which meant that for future studies, we could use the staffs' reports of caring behaviors to approximate the patients' report of caring behaviors. This should be done with caution, however, since the r-value is less than .80. It is reasonable to assert that a correlation of .80 or greater indicates the report of caring is the same, but considering the r-value is only .486, the staffs' report of caring should be used only as an approximation. These findings are noted in Figure 10.2.

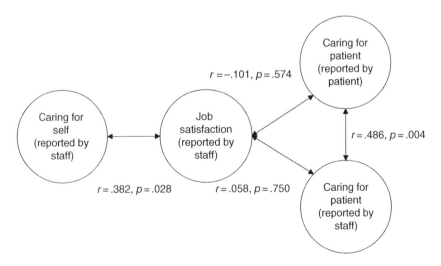

Figure 10.2 Results from testing Model 1.

Respecifying the Model

Much of the discussion after we tested Model 1 centered on the fact that the first model showed no relationship—or even showed a slight negative relationship—between job satisfaction and caring behaviors as perceived by the patient and provider. It was hypothesized that this counterintuitive finding was due to a missing variable in the model, rather than being an accurate finding. After discussion of the results, we decided to study a second model that would provide deeper insight into whether recent research was correct in suggesting that "clarity" is an important factor in job satisfaction and thus may impact the enacting of caring behaviors of staff toward patients (Nelson, Nichols, & Wahl, 2017). Felgen asserts that clarity of self, clarity of role, and clarity of system are important variables in providing patient care, as they help staff members to be innovative and courageously challenge the status quo (Felgen & Nelson, 2016).

Testing Model 2

Model 2 was devised to help examine whether clarity of self, role, and/or system influenced job satisfaction. Testing of Model 2 occurred two years after the initial study. The study of Model 2 included 161 respondents from 6 facilities, all using the same methods used to study Model 1, except for the addition of a measurement of the 3-facet construct of clarity. Additionally, this time we would use regression equations, since we were now measuring at the individual level, not the unit level. We also excluded reports of caring by the patient since we still did not have dyads of patients and staff members. However, since testing of Model 1 showed that patients and staff members were reporting caring approximately the same, it was deemed sufficient to measure caring as reported by staff members.

Along with new information from the literature on the effectiveness of clarity, results from this respecified model (Figure 10.3) provided more insight into what enhances the enactment of caring behaviors between patients and staff members.

Note that the first part of the model hypothesizes that self-care will predict clarity of system, role, and self, and the second part hypothesizes that clinicians who experience clarity of

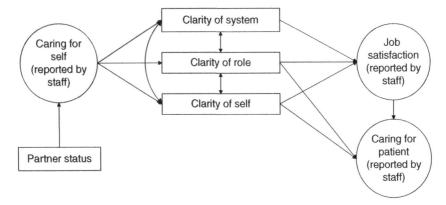

Figure 10.3   Model 2, to test relationship of clarity to job satisfaction and quality of patient care.

system, role, and self will have increased job satisfaction *and* enact caring behaviors toward the patient, as reported by staff. Note also that the demographic variable of "partner status" was added to Model 2. Findings related to this variable are reported later in this chapter.

Results from the Study of Model 2

Results from Model 1 had been generated using Pearson's correlation which revealed a simple relationship between caring for self and job satisfaction: as caring for self increased, job satisfaction also increased. But no relationship was found between job satisfaction and the enactment of caring behaviors. In our study of Model 2, we used regression analysis, which is a method of analysis that goes beyond showing a simple relationship between two variables, to showing, in this case, whether caring for self *predicts* job satisfaction and if job satisfaction *predicts* the enactment of caring behaviors as perceived by staff.

The regression analysis used in Model 2 revealed that caring for self predicts 31.1% of clarity of role, 15.7% of clarity of system, and 15.8% of clarity of self. Testing of Model 2 also revealed that clarity predicted job satisfaction, with clarity of system predicting 42.9%, clarity of role predicting 3.8%, and clarity of self predicting 2.7%. This suggested that staff members who understood their role, and how the system works, and who had self-awareness experienced greater job satisfaction. Additionally, job

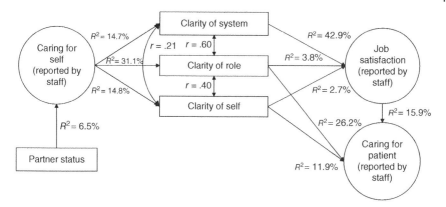

Figure 10.4 Results of Model 2 to test relationship of clarity to job satisfaction and quality of patient care.

satisfaction in Model 2 predicted 15.9% of enacting caring behaviors as perceived by staff, meaning that staff members who are expressing satisfaction with the social and technical aspects of their job perceive that they are more likely to enact caring behaviors in patient care. It is important to remember here that job satisfaction in this study is not defined as one being "happy or sad" on the job, but as "the feeling derived from perceiving that the social and technical aspects of the work environment are sufficient to perform the job" (Nelson, 2013, p. 23).

After the study of Model 2, we saw that clarity did, in fact, predict both job satisfaction and the enactment of caring behaviors (see Figure 10.4).

Interestingly, clarity of role and clarity of self also had a direct impact on the experience of enacting caring behaviors as perceived by staff, predicting 26.2% and 11.9% respectively. If clarity explains almost 50% of what predicts job satisfaction, and job satisfaction explains almost 40% of what predicts enactment of caring behaviors, it can be proposed that clarity of self, role, and system are valuable conditions to cultivate.

The Self-Care Factor

Model 2, like Model 1, begins with a measure for self-care. In Model 2, we see that staff members who reported taking time to care for themselves were also, largely, the staff members who reported clarity of self, role, and system. Self-care had the strongest influence on clarity of role, predicting 31.1% of what explains

Staff members who reported taking time to care for themselves were also, the staff members who reported clarity of self, role, and system.

clarity of role. Staff members who took time to care for themselves were also more likely to report higher levels of clarity of system and clarity of self, explaining 14.7% and 14.8%, respectively. This means that self-care predicts these important aspects of clarity. It could be theorized that those who take time to care for themselves have greater access to critical thinking due to decreased cortisol and stress. In sum, it appears that staff members who perform self-care, think more clearly; thus, there is an imperative for leaders to create environments where employees can perform self-care.

The only demographic factor that proved significant in Model 2 was partner status, with those who reported being married or domestically partnered conducting less self-care than those who were not partnered (single, divorced, or widowed), which explained 6.5% of what predicted self-care. It is not clear why being married or domestically partnered caused people to practice less self-care, and this provided an opportunity for deeper discussion with married and domestically partnered staff members regarding why they did not perform as much self-care.

Further Discussion

Nurses who reported the greatest frustration with every dimension of job satisfaction also had patients who reported the highest levels of caring behaviors from the staff.

Findings from Model 1, seen in Figure 10.1, suggesting that there is no relationship between job satisfaction and quality patient care, are consistent with the rather startling findings of Persky, Nelson, Watson, and Bent (2008), who discovered that nurses who reported the greatest frustration with every dimension of job satisfaction also had patients who reported the highest levels of caring behaviors from the staff. However, Persky et al. (2008) used only a single tool to measure caring (the CFS; see Appendix K), which measures caring as perceived by the patient. Our study, in contrast, used several tools from various caring theorists to measure the perception of caring, as reported by the patient (Appendix K), and thus strengthens the argument that nurses who are more frustrated with the social and technical dimensions of their job are still able to provide satisfying care as reported by the patient. It was posited in discussion of the initial interpretation of Model 1 that nurses who have clarity of self, role, and system have extremely high standards for themselves and others and "just get the job done" despite the mountain of obstacles they perceive. Hence, the model was respecified to study clarity, a key factor in the provision of quality care, as proposed by Felgen and Nelson (2016).

This finding of the negative relationship between caring as reported by the patient and nurse job satisfaction illustrates the importance of studying the Profile of Caring® as reviewed in Chapter 6. Recall the Profile of Caring is a trademarked group of variables measuring (a) clarity of role, (b) clarity of system, (c) caring for self, (d) the caring of the unit manager, and (e) job satisfaction. Chapter 6 proposed that higher Profile of Caring scores would correlate with better patient outcomes, but here we see that the five constructs within the Profile of Caring act dynamically, so a correlation between high Profile of Caring scores and better patient outcomes may not reliably happen. This dynamic behavior of the Profile of Caring may be especially true during the implementation of a framework of care; as nurses become clear in their role and system and as they learn to work within the framework of care to establish social and technical dimensions of the job, many aspects of their disposition cannot help but change. Deeper study is warranted to understand the peak effect of each of these five constructs on outcomes of care.

Summary

The discussions the data analyst had with staff members and leaders about the results were pivotal in refining how we measure, operationalize, and support caring. Model 2 shows that self-care is a predictor of job satisfaction and, by extension, of the enactment of caring behaviors by staff members. As leaders consider this relationship, questions about how they will communicate the importance of self-care to caregivers and build opportunities for self-care into the work environment are of utmost importance. While patients are the focal point of care outcomes, if clinicians do not first care for themselves, the enactment of caring behaviors as perceived by patients and families will happen less reliably.

If clinicians do not first care for themselves, the enactment of caring behaviors as perceived by patients and families will happen less reliably.

If leaders are to bring common language and application of research findings to staff members, they must first understand what the results represent in daily work. Leaders can become informed by taking an interest in the measurement process itself. As they see the refinement of models derived from staff members describing what the data means to them, leaders can understand the work to the degree necessary for them to truly be informed partners in the redesign of care delivery.

11

Pause and Flow: Using Physics to Improve the Efficiency of Workflow

Jacklyn Whitaker, Benson Kahiu, Marissa Manhart, Mary Ann Hozak, and John W. Nelson

Bejan and Zane (2012) developed constructal theory to study the laws of design in nature. They report, "For a finite-size flow system to persist in time, its configuration must evolve in such a way that provides easier access to the currents that flow through it" (2012, p. 76). Since the same can be said about design of work, Bejan and Zane's original constructal theory was adapted to study flow (and what interrupts flow) in work environments where patient care occurs. The analyst for this study, Dr. John W. Nelson, describes his adaptation of constructal theory, adding new vocabulary specific to the study of pause and flow in healthcare environments, as follows:

Types of Pause

Obvious Pause: This is easy to spot, such as someone yelling inappropriately while work is being performed, or an autocratic boss who pulls employees away from work.

Varied Pause: Some pauses vary by demographic. For example, a talkative coworker can be a pause for an introverted employee while a talkative coworker may create energy for an extroverted person.

Subtle Pause: These types of pauses are not noticeable unless careful attention is paid. For example, if the overhead music in the waiting room is soft jazz, and there is classical piano in the

Using Predictive Analytics to Improve Healthcare Outcomes, First Edition.
Edited by John W. Nelson, Jayne Felgen, and Mary Ann Hozak.
© 2021 John Wiley & Sons, Inc. Published 2021 by John Wiley & Sons, Inc.

clinical area, moving from one setting to the next may create a physical adjustment that is not bothersome, or even noticed, but it was an adjustment or pause just the same. If these pauses occur throughout the day, it is proposed that while the person may not be cognizant of the adjustments, the unconscious mind is fatigued by the multiple adjustments. The result is that spending time in the place where multiple unrecognized adjustments have occurred leaves the person feeling like "something was off today" or "there was something unpleasant about that place that I cannot identify."

Types of Flow

Obvious Flow: These are the easy-to-identify things that increase the flow of work, such as collaborative teams, supportive management, and clean, well-organized work areas.

Varied Flow: What might create flow for one person might vary by personality type or demographic. For example, some nurses working in critical care love intense clinical situations while some may find these situations to not increase flow, and may even create pause. This may be due to the person's personality or preferences.

Subtle Flow: These are aspects of work that aren't easily noticed but nonetheless enhance the experience and flow of work. For example, a subtle smile from a patient or a word of thanks from a coworker may not have obvious impact on work, but it may still have a positive impact on workflow.

Methods

The pilot for the pause and flow study used a cross-sectional research design which means all employees on the high-performing medical care unit and radiology department ($n = 194$) were invited to participate in the pause and flow study. High-performing areas of care for this study were defined as those units or departments that consistently reported high scores for outcome studies conducted by the data analyst of the organization who conducted research regarding the work environment. The radiology department had 7 small units in the department that spanned both the urban hospital and another 229-bed community hospital in the same healthcare system, while the medical unit, called R6S, was limited to the 650-bed urban care center in the healthcare system. Numbers of staff members invited from the medical and radiology units and associated hospital are noted in Table 11.1.

Leaders and all potential respondents on the pilot units were educated on the three types of pause and three types of flow. Potential respondents were provided with a 2-minute video on the types of pause and flow at the start of the study. The time was limited to 2 minutes to enable them to take a small amount of time during working hours to watch the video. Managers and unit leaders viewed a 10-minute video that included a more in-depth review of physics and constructal theory, along with the three types of pause and flow. The more in-depth review was

Table 11.1 Number of staff members by unit and campus.

Unit	Facility	Sample
R6S	Urban hospital	59
Radiology	Urban hospital	34
Rad. Sup.	Urban hospital	30
s/b Rad.	Urban hospital	2
Ultrasound	Urban hospital	16
Cat Scan	Urban hospital	13
Int. Rad.	Urban hospital	3
MRI	Urban hospital	9
Nuclear	Urban hospital	4
Total	**Urban hospital**	**170**
Radiology	Community hospital	16
Cat Scan	Community hospital	2
Ultrasound	Community hospital	3
MRI	Community hospital	1
Nuclear	Community hospital	2
Total	**Community hospital**	**24**

provided to managers to ensure they would be able to provide adequate support to staff members who might have questions about the study or questions about the pause and flow of work.

"Pause" was defined for this study as a moment that brings forth a hesitation— any moment that interrupts one's train of thinking and/or flow of work.

Simply stated, a "pause" was defined for this study as a moment that brings forth a hesitation—any moment that interrupts one's train of thinking and/or flow of work. It may be a feeling that something is not right or out of place and thus creates pause. For clinicians paying heightened attention, it may even be physically felt, like a subtle response to stress. It may be felt in the hands, the chest, behind the eyes, or elsewhere in the body, depending on individual responses to stress or frustration. It is not an additional task per se, but a pause that interrupts the flow of required tasks or other aspects of work and thus decreases their efficiency or productivity. Respondents were asked to report pause at the end of each day for one full week by responding to a short electronic survey with three items about pause. The first question asked about the most considerable pause experienced during the day and whether the pause was noticed immediately while working or whether reflecting on the day was required. Then respondents were asked about the second and third most considerable pauses along with immediate or delayed notice of the pause.

Flow was defined as something experienced during the workday that enhanced the ease, productivity, and experience of the work.

Flow was defined as something experienced during the workday that enhanced the ease, productivity, and experience of the work. Flow was identified as generally more difficult to specify. Examples included a compliment from a patient that created a good feeling, or a clean work environment that helped work performance because everything was in its proper place. Respondents were asked to report flow at the end of each day for one full week by responding to a short electronic survey with three items about flow. The first question asked about the most considerable flow experienced during the day and whether the flow was noticed immediately while working or whether reflecting on the day was required. Then respondents were asked about the second and third most considerable moments of flow along with immediate or delayed notice of flow.

The study occurred after approval from the ethics review board of the hospital. Employees could respond to the survey at any time during the day. There was an option at the bottom of the electronic survey to "save and return" as often as they liked if the employee wanted to have more time to respond. They were asked to respond every day they worked during the 7 days the study was conducted. A duration of 1 week was selected as this was deemed sufficient by the lead analyst of this study to identify the pause and flow of each unit. Submission of the survey was considered consent to use participants' responses in this study and to present results in aggregate form to ensure that participants were not identified.

The responses from the participants were examined by two analysts trained in qualitative methods. Items were examined for agreement on identified themes. Only those few responses that were not agreed upon were discussed between the two analysts until consensus was reached for appropriate coding. Responses where consensus could not be reached were coded as "miscellaneous."

Sample Size and Response Rates

The medical care unit had variable scheduled staff, depending on daily needs. There were 59 total staff members with 11–24 staff members scheduled each day, which calculated to 161 possible responses for the 7-day study. There were 135 staff members on the 7 radiology units with 89 scheduled on weekdays and 30 scheduled on weekends, which calculated to 505 possible responses on the radiology units.

There were 141 surveys completed out of a possible 161 on the medical unit (a 86% response rate) and 151 surveys completed of a possible 505 on the radiology units (a 30% response rate). A total of 292 surveys were submitted.

What We Learned About Pause

For the first question, asking respondents to identify the most considerable pause, there were 292 responses across all participating units. For the second most considerable pause, there were 281 responses. For the third most considerable pause, there were 256 responses.

Examination of themes for the most considerable pause revealed that the medical unit (referred to as R6S) reported "interruption of work" as the most frequent incident of pause, the radiology department reported that the telephone was the most considerable pause, and the ultrasound department reported that transport delay was the most considerable pause. The other units had more varied reports of pause across reporting staff. All responses to the most considerable pause are noted in Figure 11.1.

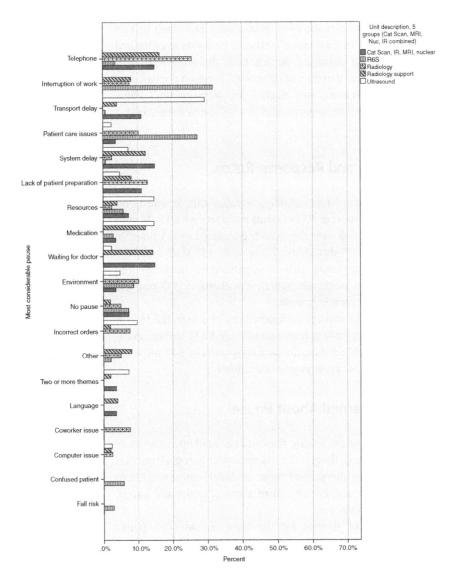

Figure 11.1 Most considerable pauses, by percentage of pause themed by unit (each unit calculated to 100%).

It was revealed that the second most considerable pauses identified were "no pause," followed by lack of patient preparation and incomplete orders. See Figure 11.2 to discover more about what caused the second most considerable pauses.

The third most considerable pauses, after "no pause," included lack of resources, incorrect orders, and problems in the physical work environment. See Figure 11.3 to discover more about what caused the third most considerable pauses.

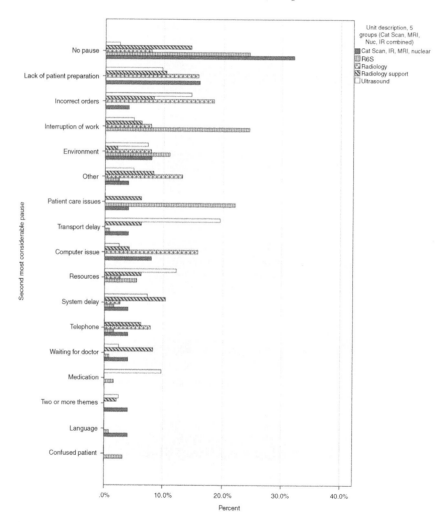

Figure 11.2 Second most considerable pauses, by percentage of pause themed by unit (each unit calculated to 100%).

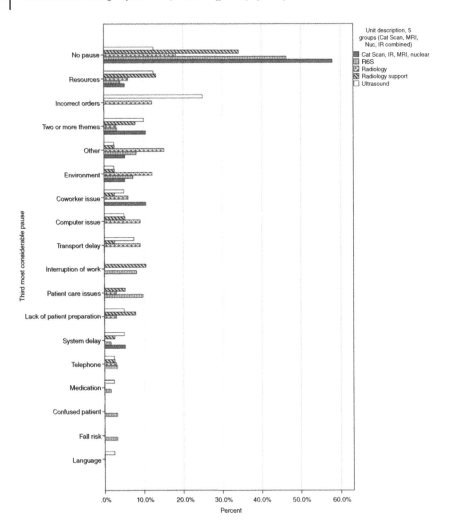

Figure 11.3 Third most considerable pauses, by percentage of pause theme by unit (each unit calculated to 100%).

What We Learned About Flow

"Good coworker relationship" was the most frequent predictor of flow across the participating units.

Examination of frequency of themes revealed that "good coworker relationship" was the most frequent theme across the participating units. Receipt of an affirming comment from a patient, family member, or visitor was the second most frequent theme causing the most considerable flow. To discover more about what created the most flow, see Figure 11.4.

In our study of the second most considerable flow, "none" was the most frequent answer given, followed very closely by good coworker relationships and correct orders/paperwork. A distant but significant third, is timeliness, and it is also notable

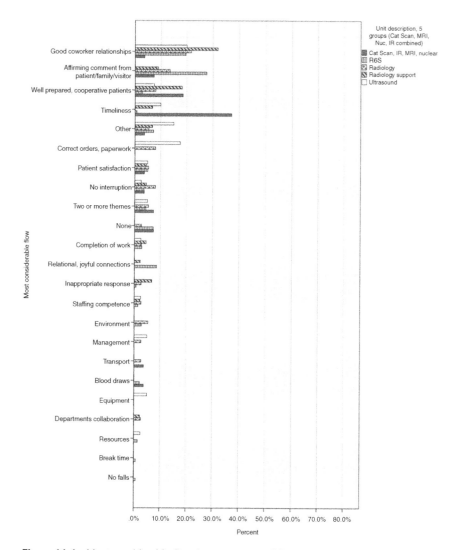

Figure 11.4 Most considerable flow, by percentage of flow themed by unit (each unit calculated to 100%).

that on the surgical unit, "affirming comments from patient/ family/visitor" was a frequent response. See Figure 11.5 for more information on what caused flow.

In our study of the third most considerable flow, again, "none" was the most frequent answer given, followed by good coworker relationships; correct orders/paperwork; and well-prepared, cooperative patients. See Figure 11.6 to see more about what caused flow for our respondents.

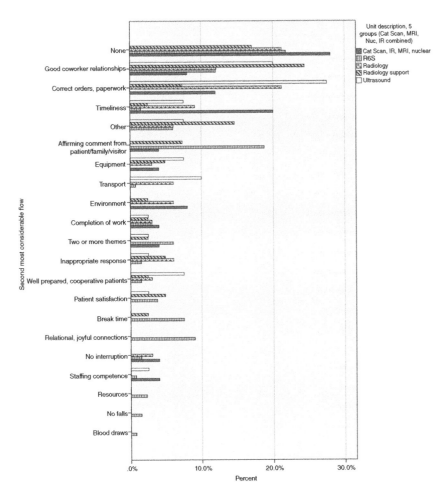

Figure 11.5 Second most considerable flow, percentage of flow themed by unit (each unit calculated to 100%).

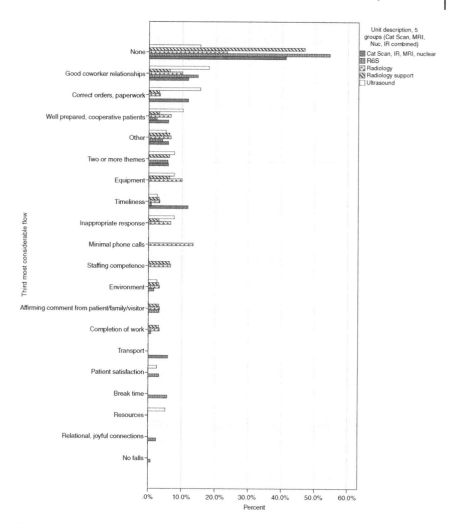

Figure 11.6   Third most considerable flow, by percentage of flow theme by unit (each unit calculated to 100%).

Recall or Reflect?

You will recall that along with asking participants to identify incidents of pause and incidents of flow, we asked whether they could easily recall their moments of pause or flow or if they had to reflect in order to recall those moments. Our aim with this question was to determine which instances—pause or flow—were more impactful. We discovered that moments of pause were remembered more often and more easily than moments of flow. For details about participant responses to this section of the survey, see Appendix L.

Application of Results to Operations

Use of the pause and flow study on the pilot units helped identify areas of concern. Staff members found the surveys short and easy to respond to, and the resultant report helped staff members see whether their experience of work was similar to or different from that of other workers on the unit. It was also helpful for managers to see the most frequent responses reported by staff members from their respective units. Understanding the most common impediments and facilitators of work helped managers prioritize efforts for operational refinement. Because the study proved quick, interesting, and useful, it was replicated across all other units and used periodically thereafter to identify specific areas of concern.

The pause and flow study not only helped identify broken systems—it helped identify areas of concern for professional practice.

The pause and flow study not only helped identify broken systems—it helped identify areas of concern for professional practice. For example, one of the most common responses for pause was being "interrupted" by patients for care, a typical response being, "the patient stopped me to ask a question." This helped leaders understand that not all staff members were clear about their role. It appeared that some staff members were so focused on tasks that they viewed the reason for the very existence of their professional role, the patient, as a pause in their work. This was surprising, and it is a testimonial as to why this method of analytics is so useful.

Reflections from the Medical Unit—R6S

This study provided valuable lessons during both the data collection period and during examination of the results. People in different roles within R6S also learned about unique aspects of their flow of work; this was true for staff members and unit leaders. This section will review how our pause and flow study provided lessons at different stages and in different ways.

Lessons from Engaging in the Study Itself

One of the most important insights R6S learned in the pause and flow pilot study was that interruptions occurred surprisingly frequently throughout the day. These interruptions had become so embedded as a part of daily work, however, that the pauses were considered normal. Because the pauses were seen as normal (and unavoidable), the staff of R6S were not aware that they could be more productive if they somehow eliminated these pauses. Staff had long accepted things such as pharmacy not dispensing all of the patient's medications on time (or at all) as a part of the daily routine. The pause of medications not being dispensed as expected resulted in the nurse calling pharmacy and then leaving the unit to go get the needed medication from the pharmacy department. It was a big pause, it happened often, and it was avoidable.

Staff had long accepted things such as pharmacy not dispensing all of the patient's medications on time (or at all) as a part of the daily routine.

Another common example of an interruption was staff members accepting phone calls from patients' family members during the change-of-shift hand-off report which occurs at the patient's bedside. If the phone rang during hand-off, the hand-off report stopped until the phone call was completed. It had also become acceptable to have at least one or two blood pressure machines not working at all times, which required staff members to share— sometimes having to leave a patient to retrieve the proper equipment. While these pauses in work had implications for efficiency, no one had realized how these all-too-common pauses had implications for patient safety. As the pauses and flows were identified and examined, it became clear that unit staff members were spending a significant amount of time being diverted from patient care.

This widespread acceptance of unnecessary pauses became clear to unit leaders during the data collection period of this study. It was the resource nurse who had assisted staff members during the study, who realized that most members of the staff believed the pauses were a normal, unavoidable part of the work. The resource nurse helped staff members understand that the elements of dissatisfaction they were experiencing in the work were actually pauses that were not part of the process of work, but a broken part of the system. Staff members never brought these frustrations to the advisory committee or the nurse manager before, because no one had perceived them as problems that could or should be solved.

Staff members never brought these frustrations to the advisory committee or the nurse manager before, because no one had perceived them as problems that could or should be solved.

A Lesson from the Results of the Study

The data the study produced helped the unit advisory council, which included both management and staff members, to dissect what was normal and necessary and what was broken in the process of work. Some of the pauses identified by staff members included lack of available medication, lack of supplies, phone interruptions, and other pauses identified in Figures 11.1–11.3.

Once these pauses were formally identified, the resource nurse began to mitigate these pauses as much as possible. For example, the nurse-to-nurse hand-off at change of shift is a critical time to transfer patient information from the nurse going off duty to the nurse just arriving for work. Proper transfer of information from nurse to nurse has implications for continuity of the patient's plan of care, patient experience, and even patient safety. Prior to the pause and flow study, nurses were interrupted for phone calls during bedside hand-offs. This interruption was identified as a pause. However, the importance of communication during nurse hand-off does not negate the importance of supporting families and responding to their questions and concerns; both were of high importance. After the study completion and discussion of the results, the advisory council came up with a plan to meet both needs. Whenever a patient's family member would call during the hand-off period, the resource nurse would speak with the patient's family member to provide guidance on best times to call for updates about their family member and provide reassurance about the patient's status and the certainty that they would be able to connect with the nurse during the provided hours. This helped the nurses resume safe hand-off communication while still providing time for the nurse to speak with the family member after the important information was communicated. Additionally, after hand-off, the nurse was likely to have even more pertinent information to communicate to family members.

The most important aspect of the pause and flow process was that staff members became adept at identifying daily aspects of their own work routines that were pauses in their workflow, and they had new insight into what was unavoidable and what could be fixed. Through this identification, staff members were able to plan their day in ways that would intentionally reduce pause and increase flow.

What the Manager Learned About Her Own Practice

The manager of this medical unit reported that the pause and flow process helped her clearly see work pauses that were keeping staff members from completing their jobs on both the 8- and 12-hour shifts. At one point, the unit manager overheard a patient care attendant (PCA) saying that the nurse manager had caused a pause by stopping what the PCA was doing with a patient to remind her to attend a staff meeting. This moment of realization allowed the manager to identify that her own interactions with staff members were sometimes causing the staff to experience unnecessary pauses. Subsequently, the manager reassessed how she communicated with staff members to not only ensure that she was facilitating flow, but to be much more intentional about not assigning tasks that created pause. Being more aware of pause and flow, she began more proactively communicating with the unit staff as well as professionals from other disciplines who were working on the unit to understand what pauses they were experiencing during their workday. A real-time assessment of pause and flow on her unit enabled her to intervene, in real time, to remove pauses and facilitate flow.

The nurse managers' own interactions with staff members were sometimes causing the staff to experience unnecessary pauses.

Changes Made on the Medical Unit Because of the Study

Primary improvement processes that were implemented on the medical unit after the pause and flow study included using patients' folders, an Out-of-Bed project, and medication administration. The patient folders created a central location for all educational material for patients, a standardized location for each patient's belongings list, and an accountability process to ensure that patients consistently received education material regarding medications and side effects. The Out-of-Bed project promoted lung expansion, flow of oxygen, prevention of skin breakdown, prevention of blood clots, strengthening of muscle, and prevention of falls, all of which supported the organization's high reliability initiatives.[1] Patients who would normally be bedbound because of various physical limitations, were now assisted to and from bed via two-person assist or appropriate

---

1 High reliability organizations use consistent processes to avoid catastrophic events (Kerfoot, 2006).

equipment. The medication administration project's goal was to create an environment for nurses to provide focused, uninterrupted medication administration and education which was rolled out on R6S as well as R6S's sister unit, R4N. The implementation of these pause-reducing projects has enhanced patient outcomes such as reduction of fall rates, increase of patient satisfaction, and decrease of nurse turnover on R4N, also demonstrating that reducing work pause while increasing workflow improves nurse job satisfaction.

The three issues that were prioritized to most enhance outcomes were:

- Pharmacy not dispensing patient's medications on time or at all,
- Telephone interruptions of the nurses, and
- Lack of functional equipment (BP machines, O2 regulators, bed alarms).

Interventions designed to address each of these issues can be found in Appendix M.

Analyzing Pause and Flow of Work as a Method of Quality Improvement

The success of the pause and flow project, in the pilot and again on all units, was significant. It was found to be an exciting way to address efficiency because it used theming of comments from staff members to identify the most pressing issues of workflow. It was engaging for staff to be rid of the pauses and useful for leaders because staff members were engaged in making changes for more efficiency and better outcomes. As a result, pause and flow analysis became a standard method of process improvement in the organization.

Summary and Next Steps

By tapping into the most enjoyable aspects of work, as well as the frustrations that had become a part of the daily routine, the study brought along with it the passion of employees to support change. The discovery of issues that were causing staff members to pause helped identify the frustrations of staff. This insight helped leaders take targeted measures to make the work experience more enjoyable while also improving efficiencies and productivity. This process could be used prior to a Lean project, to help engage staff members in identifying issues in need of refinement using a process they believe in and enjoy.

By tapping into the most enjoyable aspects of work, as well as the frustrations that had become a part of the daily routine, the study brought along with it the passion of employees to support change.

A very compelling follow-up project would be to return to the idea of whether it was the frustration of pause or the lift that comes with flow that had the bigger impact on staff members. Imagine what you could do if you had data on whether it was the elimination of pauses or the increase of flow experiences that was the bigger predictor of turnover or employee retention and which had the bigger impact on costs! Cleary there is more to discover in the study of pause and flow.

12

Lessons Learned While Pursuing CLABSI Reduction

Ana Esteban, Sebin Vadasserril, Marissa Manhart, Mary Ann Hozak, and John W. Nelson

Throughout 2015 the quality improvement team at our 650-bed urban hospital in the Northeastern United States developed a series of predictive models designed to proactively manage outcomes such as central line-associated blood stream infections (CLABSI). Our hospital had fully implemented Relationship-Based Care® and all of its principles (Creative Health Care Management, 2017; Koloroutis, 2004). We began with a performance improvement project to build a model specified for CLABSI reduction for the context of this hospital and its processes of clinical care. We spent all of 2015 introducing the initiative to staff, getting approval from appropriate committees, and sequencing the steps for implementation in 2016.

Development of a Specified Model of Measurement for Prevention of CLABSI

The primary source to build the initial model of measurement was the guidelines published by the Centers for Disease Control (CDC) (O'Grady et al., 2011). The CDC document ranked the guidelines for central line care based on level of the evidence as follows:

- The highest level of evidence, listed as 1A, indicated, "strongly recommended for implementation and strongly supported by well-designed experimental, clinical, or epidemiologic studies" (p. 8).

Using Predictive Analytics to Improve Healthcare Outcomes, First Edition.
Edited by John W. Nelson, Jayne Felgen, and Mary Ann Hozak.
© 2021 John Wiley & Sons, Inc. Published 2021 by John Wiley & Sons, Inc.

- The second highest level of evidence, listed as 1B, indicated, "strongly recommended for implementation and supported by some experimental, clinical, or epidemiologic studies and a strong theoretical rationale; or an accepted practice (e.g. aseptic technique), support by limited evidence" (p. 8).

These 2 levels of evidence, and data from the existing audit of central line care, were used to develop the first 89-item model of measurement. The graphic to depict the level of evidence and what was being and could be collected is found in Figure 12.1. A key naming every variable in Figure 12.1 is found in Table 12.1.

It is important to note that only the shaded circles in the model represent variables for which data was collected by the organization's existing CLABSI measurement tool and subsequently provided a starting point for this study. These variables were collected from the existing audit and deemed the best way to get the study and process going. Notice also, in Figure 12.1, that the (1A) variables are encircled in one color and the (1B) variables are encircled in another color. Use of this model depicted in this way helped us to target data that aligned with the evidence. The circles that have no color in them were not collected and were not considered for this initial study. We wanted to collect all variables, but it was found that the data was either not accessible for the study or was reported by the vendor(s) to be too costly or time-consuming to provide for this study.

The description of each of these variables has been listed in Table 12.1, along with the level of evidence and whether the variable was available for electronic (versus manual) data collection.

The ultimate goal of the study was to interface all software in the organization that contained data related to CLABSI and then use predictive analytics to proactively manage risk for CLABSI in real time. Such interoperability would also free up time for quality improvement staff to work on model validation and refinement instead of manually collecting data.

The first analysis, which included only those items that were in the existing audit for central line care (the shaded circles noted in Figure 12.1), was conducted in June 2016. Data was collected from 10 inpatient units from April to July 2016. We used only these items because the tool was already developed

Figure 12.1 Specified model to study central-line associated blood stream infections (CLABSI).

and it included variables of interest relevant to the realities of the current team. Despite the tool being the same one they already knew, the process and approach would be different. The team was interested in moving beyond the study of frequencies of CLABSI to studying what variables would predict CLABSI. The data scientist taught the staff that the first step in generating data that would inform not just frequency of CLABSI but what *predicted* CLABSI, is to collect and measure only valid, reliable data. Flawed data, or data collected with a flawed process, would not reveal what predicts CLABSI.

Table 12.1 Variable descriptions from specified model of measurement for prevention of CLABSI.

Variable number	Variable description	Evidence level	Possible to collect electronically
1	Interdisciplinary QI process	**1B**	
2	Unit		x
3	Service line		x
4	Room number		x
5	Medical record number (MRN)		x
6	Account number (new number each admit)		x
7	RN reviewer (person conducting audit)		
8	Date of review (audit by RN reviewer)		
9	Name of RN caregiver during central line (CL) audit		x
10	Shift of RN caregiver day of audit (day/night)		x
11	Name of intensivist group day of audit		x
12	Does patient have a CL (yes/no)		x
13	Type of line (TLC, PICC, etc.)		x
14	Location of the line	**1A**	x
15	Daily assess need for CL	**1A**	x
16	Number of CLs		x
17	Intravenous (IV) bags and tubing labeled and dated (yes/no)		
18	No prophylaxis antibiotic before insert or use of CL	**1B**	
19	IV bag/tub changed 72 hr (yes/no)		
20	Chlorhex. Gluc impreg sponge on CL (yes/no)	**1A**	
21	CL stabilization device on (yes/no)		
22	CL insertion date/time documented? (yes/no)		
23	Central line dressing protocol followed	**1A**	x

Table 12.1 (Continued)

Variable number	Variable description	Evidence level	Possible to collect electronically
24	Central line ports free of dried blood? (yes/no)		
25	Dressing change date documented? (yes/no)		
26	Facility		
27	Staff education in prevention of CLABSI	1A	x
28	Assess periodically knowledge in CL care	1A	x
29	Assess periodically CL care (e.g. survey #328, CL care audit)	1A	
30	Line competence	1A	x
31	RN patient ratio	1B	x
32	Number of pool nurses	1B	x
33	Level of staff experience in line care	1B	
34	Do not use steel needles for fluid administration (peripheral and midline catheters only)	1A	
35	Do not use short peripheral catheters if care > 6 days (peripheral and midline catheters only)		
36	Palpate site daily through dressing to evaluate if tender (peripheral and midline catheters only)		
37	Remove peripheral lines if s/s of infection of line not working (peripheral and midline catheters only)	1B	
38	Use subclavian site over femoral or jugular for central line insert	1B	
39	Avoid subclavian site for hemodialysis patients	1A	x
40	Diagnosis		x
41	Do not use central venous catheter (CVC) for permanent dialysis access	1A	x
42	CL ultrasound guidance (by trained staff) preferred	1B	

(*Continued*)

Table 12.1 (Continued)

Variable number	Variable description	Evidence level	Possible to collect electronically
43	Minimize use of lumens in CVC	1B	
44	Replace CL within 48 hr if placed in emergency	1B	x
45	Hand wash before and after touching CL site	1B	
46	Aseptic technique for insert and care of CL	1B	
47	CL port changed every 7 days or as necessary		x
48	IV tubings covered with sterile caps		
49	Removal of CL documented		x
50	Location where CL inserted (i.e. unit)		x
51	Primary IV tubing change every 96 hr? (yes/no)		x
52	Alcohol impregnated caps in place? (yes/no)		
53	CL dressing is clean, dry, and intact	1B	
54	Sterile gauze or transparent dressing to cover site	1A	x
55	Monitoring devices	1A	x
56	Administration sets	1A	x
57	Needleless IV catheter system	1A	
58	Monitor ongoing use of catheter and associated risk	1B	
59	Sterile gloves worn when insert arterial line, CL, midline catheter	1A	
60	Use max. sterile barriers (cap, gown, precautions)	1B	
61	Prepare skin using >0.5% chlorhexidine for insertion and dressing change	1A	
62	Site of antiseptic scrubbing dry prior to CL insertion	1B	
63	Use povidone iodine or bacitracin ointment at the hemodialysis catheter exit after catheter insertion and at the end of each dialysis session	1B	

Table 12.1 (Continued)

Variable number	Variable description	Evidence level	Possible to collect electronically
64	No need to replace peripheral catheters more frequently than 72–96 hr to reduce risk of infection and phlebitis in adults	1B	
65	Do not routinely replace CVCs, PICCs, hemodialysis catheters, or pulmonary artery catheters to prevent catheter-related infections	1B	
66	Do not use guidewire exchanges routinely for non-tunneled catheters to prevent infection	1B	
67	Do not use guidewire exchanges to replace a non-tunneled catheter suspected of infection	1B	
68	Use a guidewire exchange to replace a malfunctioning non-tunneled catheter if no evidence of infection is present	1B	
69	In adults, use of the radial, brachial, or dorsalis pedis sites is preferred over the femoral or axillary sites of insertion to reduce the risk of infection	1B	
70	A minimum of a cap, mask, sterile gloves and a small sterile fenestrated drape should be used during peripheral arterial catheter insertion	1B	
71	Use disposable, rather than reusable, transducer assemblies when possible	1B	
72	Replace disposable or reusable transducers at 96 hr intervals. Replace other components of the system (including the tubing, continuous-flush device, and flush solution) at the time the transducer is replaced	1B	
73	Keep all components of the pressure monitoring system (including calibration devices and flush solution) sterile	1A	

(Continued)

Table 12.1   (Continued)

Variable number	Variable description	Evidence level	Possible to collect electronically
74	When the pressure monitoring system is accessed through a diaphragm, rather than a stopcock, scrub the diaphragm with an appropriate antiseptic before accessing the system	1A	
75	Do not administer dextrose-containing solutions or parenteral nutrition fluids through the pressure monitoring circuit	1A	
76	Sterilize reusable transducers according to the manufacturers' instructions if the use of disposable transducers is not feasible	1A	
77	In patients not receiving blood, blood products or fat emulsions, replace administration sets that are continuously used, including secondary sets and add-on devices, no more frequently than at 96 hr intervals, [177] but at least every 7 days	1A	
78	Replace tubing used to administer blood, blood products, or fat emulsions (those combined with amino acids and glucose in a 3-in-1 admixture or infused separately) within 24 hr of initiating the infusion	1B	
79	Replace tubing used to administer propofol infusions every 6 or 12 hr, when the vial is changed, per the manufacturer's recommendation	1A	
80	Other (to be determined)		

There were 258 complete assessments of the central line care from 159 patients. This first period of data collection was conducted by one performance and safety improvement (PSI) staff member as the program was being piloted. Data for the remaining four periods was collected by 22 RNs. There were 5,986 total line assessments made across 5 models of measurement in 2,848 patients over the study period (see Table 12.2).

Table 12.2 Frequency of audits and numbers of patients and line assessments.

Study	Dates	Number of patients	Number of line assessment
Model 1 (existing audit)	January 1, 2015–March 30, 2016	NA	NA
Model 2	April 15, 2016–July 1, 2016	159	258
Model 3	July 18, 2016–August 30, 2016	388	1,039
Model 4	September 1, 2016–August 31, 2017	1,120	2,215
Model 5	October 31, 2017–March 31, 2018	1,181	2,474
Total		2,848	5,986

First Lesson Learned: Quality Data Collection Requires Well-Trained Data Collectors

The initial audit of central lines was designed to examine the extent to which each of the items from the audit predicted CLABSI. The items related to central line care performed by the RNs in clinical care were of particular interest, as it was hypothesized that the more closely the RN followed the hospital protocol for central line care, the lower the infection rate would be. To calculate the "protocol compliance score," all items for the protocol for central line care in the audit were summed and examined in relationship to CLABSI. Thus, it was hypothesized that the higher the compliance score was, the lower the infection rate would be.

Results of the initial analysis were surprising, as it was found that there was no relationship at all between compliance scores and infection rate. This was curious, so the compliance score itself was examined using regression analysis, and it was found that the compliance score related most strongly with the auditor, with "who the auditor was" predicting 20% of the variance of protocol compliance. This finding was statistically significant, and it meant protocol compliance was not predicted by the

Results of the initial analysis were surprising, as it was found that there was no relationship at all between compliance scores and infection rate.

nurse who carried out the protocol for central line care, but rather by the auditor.

It was found through conversations with auditors that the data was not being collected in a consistent manner. Auditors who recorded perfect protocol compliance scores for the nurses they were auditing revealed that they had warned the nurses when the audit would occur, would return to the clinical nurse if the nurse had not had a chance to tend to the central line protocol, or were "kind" to the clinical nurse because the compliance score of the clinical nurse was part of the annual review and their subsequent financial reimbursement. Thus, true practice was not being captured by the auditors reporting perfect compliance scores. This finding provided a lot of insight for the quality team leaders who oversaw the performance and safety improvement program and the work performed by the auditors. Most of the quality team leaders also conducted audits themselves.

True practice was not being captured by the auditors reporting perfect compliance scores.

After leaders of the performance and safety improvement program became clear on the methods of some of the auditors and the subsequent poor inter-rater reliability (which checks whether everyone is auditing the same way), a formal training program for the auditors was developed. The program was designed to reinforce the aims of:

1) Studying predictors of central line associated infections,
2) Clarifying the purpose of each item in the audit by asking auditors to explain their interpretation of each item,
3) Discussing the methods for data collection to ensure consistency across all auditors, and
4) Facilitating discussions with auditors regarding operational challenges in conducting audits.

The training program included classes on aspects of data collection, including clarity of verbiage of each audit item and inter-rater reliability. The class concluded with a test. The goal of the class was to improve the inter-rater reliability of how audits were conducted and thus was called the Inter-Rater Reliability Class. Our desire was to reduce "who the auditor was" as a predictor of compliance. The class curriculum was developed by the quality leaders and the data analyst, who was trained in developing assessment items that could secure the information desired to gauge each auditor's ability to properly assess nurses for compliance with protocols for central line care.

The goal of the class was to improve the inter-rater reliability of how audits were conducted.

To identify the current knowledge of the auditors, the test was administered at the beginning of the class, before any instruction was provided, and again three weeks after the class was attended. The test assessed what auditors should record as "compliant with protocol" and what should be recorded as "non-compliant with protocol." The test included vignettes based on actual clinical scenarios encountered by auditors. Each vignette allowed auditors to demonstrate clinical judgment used to score care in complex situations.

What follows is a sample vignette from the test and subsequent multiple choice options for auditors taking the test. There was one vignette for each aspect assessed for compliance.

Sample Vignette: You are auditing the central line on a patient with a Triple Lumen Catheter. The patient does not have an alcohol impregnated disinfecting port protector on the distal port. The RN states that the unit does not have any port protectors stocked in the supply closet, and you have heard there is a shortage throughout the hospital. You know from many previous audits that this nurse would be compliant if the supplies were available.

Question Associated with Vignette: Based on this clinical scenario, how would you answer the following question: Alcohol impregnated disinfecting port protectors are in place (including the port/s of the peripheral lines)? Options for response included (a) Yes, (b) No, (c) Skip this question.

The correct answer is (b). Regardless of the nurse's history of compliance and the shortage of port protectors, this aspect of the hospital acquired infection (HAI) prevention protocol is not met in this scenario. The patient is at risk of infection since this is not being done, even though the reason is out of the nurse's control. The auditor must be objective in this and every other scenario. If this were a real-life situation, the auditor would then follow up with the staff member and manager later to resolve the issue.

Some of the vignettes were accompanied by photos of the central line dressing or a component of the central line addressed in the vignette. In the baseline test, taken before the class began, 70% of the auditors correctly identified compliance with the protocol. However, only 37% identified noncompliant components in the photographs.

The class was formatted into three parts:

1) Reviewing study aims and the auditor's role in gathering meaningful data,
2) Clarification and review of survey questions, and
3) Discussion with auditors about barriers to quality data gathering.

The class had ample time for discussion about data collection, and the presentation left substantial time for an open forum with discussion of barriers not only to compliance with the HAI prevention protocols, but also of their own obstacles when auditing their peers. We expected to hear reports of hostility from staff members, which the auditors stated was minimal. Auditors self-reported that the main influence on their previous auditing decisions was a perception that the results would be used in punitive measures. Some auditors expressed that they were uncomfortable including primary nurses' names in their audit reports. It was also reported that some of the audit items were not as clear as they could have been, indicating some lack of reliability in the audit itself.

At the end of the initial year of the program, noticeable and significant results were noted in the incidence of CLABSIs (number of CLABSI infections/1,000 device days) throughout the division: The program was able to reduce 24 CLABSIs in 2016 compared to 2015. Using cost analysis from Zimlichman et al. (2013), this represents a savings, after subtracting the cost of the nurse auditors and the data analyst, in the amount of $956,301 for the organization for 2016. The most important result of this CLABSI reduction study, however, is that there were 24 fewer patients and families adversely affected.

This program reduced CLABSIs by 24 in 2016, with a cost savings of $956,301.

Other Lessons Learned

Pursuit of rigorous data collection with resultant sound data helped the team advance the conversation. There were several meaningful discoveries made.

Lesson 1: Auditors Can and Should Mentor Staff Members

Leaders of the performance, safety, and improvement (PSI) program, who also did audits of central line care, used deep knowledge of proper line care to mentor staff. When rounding to audit central line care, nurses would speak to the auditor with apprehension with comments such as, "Don't look at my line," "I'm sorry, I didn't get to it yet, but I will," or "I just came on, and it was like this." The auditors started using the audits as teaching moments. After the audit was complete, the auditor would show the nurse what was done well and what needed correction. The auditor would then change the line with the nurse and demonstrate proper technique. This helped the nurse learn that the auditors were not checking lines to be punitive, but to support professional development and consistent quality care.

Lesson 2: Risks Identified in the Literature May Not Match Your Reality

Our analysis of predictors helped us identify risk factors that had not initially been considered. For example, the femoral lines in our study had the lowest rates of infection, while the literature cites that these have the highest rates. Discussion with staff members revealed that when a femoral line was placed, there was urgency to remove it as quickly as possible, due to what was known in the literature about its high rates of infection. This appeared to work the opposite for peripherally inserted central catheter (PICC) lines, which were known to be the safest in the literature but had the highest infection rates in our data. Staff revealed that they felt PICC lines were so safe that there was not the same urgency to remove them as there was with the femoral lines. It appears that evidence-based practice was relied upon too strongly—and to the exclusion of considering context of operations.

Femoral lines in our study had the lowest rates of infection, while the literature cites that these have the highest rates.

Lesson 3: Practice Errors, Beyond What You Are Looking for, May Be Discovered

Using different methods of data analyses while carefully applying the scientific method made us feel more confident in our results, which then helped us move into action regarding the operations found to be impacting the outcomes. For example, when a unit was finding an increase in CLABSI with data that was validated across datasets, there was a discussion about what changed on the unit at the time of the increase. In this instance of increased infections, the only thing that changed was the laboratory staff members who were collecting blood samples, and all of them were new to the organization. After discussion with the laboratory staff about methods to collect blood samples, it was identified that the method demonstrated was contaminating the blood samples. Thus, the rise was due to contaminants and not true infections, so we retrained laboratory staff members in proper methods of collecting blood samples. Here again, our confidence in the data helped us discover what we needed to discover, through conversation about the data and associated context.

Our confidence in the data helped us discover what we needed to discover, through conversation about the data and associated context.

Lesson 4: Competence in Counting Numerator and Denominator Data Is Essential

In 2017 there was an issue with collection of denominator data (number of device days). We were benchmarking using National Healthcare Safety Network (NHSN) rates, and the rates were based on incidence of CLABSI (numerator) over device days (denominator). Accurate calculation of CLABSI rates requires understanding of what a "device" actually is, so the denominator and associated calculations are correct. The interdisciplinary team discovered that not all participants who were counting events were labeling the devices properly. For example, one of the non-clinicians who was helping with data collection was counting peripheral lines as central lines, thus falsely increasing the incident rate of CLABSI. To correct this and prevent it from happening again, an educational presentation was developed for use by anyone who was going to perform or assist with data collection on CLABSI rates. Other devices that were not clearly understood to count (or not count) as central line equipment included blood access ports, dialysis devices, PICC lines, arterial lines, and midlines. Education in all central line equipment or equipment that had been erroneously associated with central line infections became part of the annual competency day for nurses and those assisting with data collection on CLABSI rates.

Lesson 5: Leverage the Process to Advance Clinical Practice of Nurses

Dwell time of the central line was found to be a significant risk factor, as identified in the literature and this study. It was identified that private physicians, who were not part of the healthcare organization, would not remove central lines promptly; hence, some central lines remained in place longer than necessary, increasing risk for infection. It was discovered that nurses on the IV team were credentialed to remove central lines and could be deployed to intervene sooner. Having specialized IV nurses who could safely remove central lines for private physicians not only helped to decrease central line dwell time, but also enhanced the willingness of physicians to collaborate with nursing on other practice issues of care beyond CLABSI, such as addressing unnecessary blood cultures when not clinically indicated. Through collaboration with nursing leadership, intensivists, and infectious disease providers, excessive pan-culturing was reduced, thus reducing infections mislabeled as a CLABSI.

Summary and Next Steps

Developing a program to manage CLABSI is largely a function of learning how to collect data that can be trusted. Our new and growing program in management of CLABSI and other HAIs has provided insight into other areas of concern that will help specify a model to study and proactively manage incidence of CLABSI. Next steps included:

Developing a program to manage CLABSI is largely a function of learning how to collect data that can be trusted.

1) Teach performance and safety improvement (PSI) staff how to support the Relationship-Based Care concept of responsibility+authority+accountability (R+A+A) (Koloroutis & Abelson, 2017). Each staff member is responsible for their role, understands their authority to enact that role, and is accountable to conduct this role properly and professionally.

2) Continue to work toward measuring a more specified model of CLABSI that increasingly reflects the context of care while building upon the evidenced-based model that was developed using the CDC guidelines.

3) Deepen work with people in other disciplines and departments to co-create efforts of process improvement to reduce CLABSI and other HAIs.

The work we did to reduce the incidence of CLABSI and other HAIs was some of the most gratifying work any of us had ever done, and it has inspired us to do much more. The ability to target outcomes more efficiently and cost-effectively has also increased the likelihood that all of our process improvement efforts will get needed funding. When you save an organization almost a million dollars in one initiative, administration is very interested in what you might do next!

Section Three

Refining Theories to Improve Measurement

13

Theory and Model Development to Address Pain Relief by Improving Comfort

Tara Nichols and John W. Nelson

Theory is the foundation of science; it guides research design, facilitates interpretation of study results, and drives discovery of knowledge. This chapter explains how existing theories of pain management were challenged as inadequate, and a "theory of comfort" was developed, which both includes and transcends pain management (Nichols, 2018). A new model of comfort, the Nichols–Nelson Model of Comfort (NNMC) provides researchers with a new set of variables and measurement instruments that can be used to provide data about the intensity of pain and the frequency with which it is occurring, and even *why* pain is occurring or proving difficult to control.

A New Theory

The Nichols–Nelson Model of Comfort (NNMC) examines variables of pain related to the patient, the patient's context, and the healthcare system. Having a more complete understanding of what constitutes adequate pain management enables clinicians to more fully understand the frequency and intensity data related to pain, which in turn enables clinicians to co-create, with the patient, an effective plan of care for pain management. Patients' stories about their unique pain experience guided the development of this theory and its associated model and measurement instruments.

A commonly cited definition of pain is, ". . . an unpleasant sensory and emotional experience associated with actual or

Patients' stories about their unique pain experience guided the development of this theory and its associated model and measurement instruments.

Using Predictive Analytics to Improve Healthcare Outcomes, First Edition.
Edited by John W. Nelson, Jayne Felgen, and Mary Ann Hozak.
© 2021 John Wiley & Sons, Inc. Published 2021 by John Wiley & Sons, Inc.

potential tissue damage, described in terms of such damage" (Merskey & Bugduk, 1994, p. 210). This is in contrast to the definition of *comfort* proposed by Kolcaba (2003): "...the immediate state of being strengthened by having needs for relief met in four contexts of the physical, psychospiritual, social and environmental" (pp. 457–458), which is much more than the absence of pain. The NNMC embraces this definition of comfort and proposes a holistic, caring, and inclusive approach to pain management that is dependent on the quality of the patient–clinician relationship.

Developing a New Model Based on a New Theory

In order to affect a variable of interest—in this case, an outcome related to pain management—it is sometimes necessary to take a step back and evaluate whether the theory that informs our current data collection tools adequately represents the factors truly influencing the variable of interest. In this case, we step back to consider whether current measurements of pain and pain management adequately address the factors that relate to the patient, the patient's context (internally and externally), and the system in which the care occurs.

Nichols asserts that pain is modulated by factors contained in (a) the patient–caregiver relationship, (b) the beliefs of the caregiver, and (c) factors within the context of the healthcare system. Nichols collaborated with data scientist Nelson to develop a theoretical model that integrates these factors (Nichols, 2017, 2018). The model challenges both patient and clinician to think differently about pain, this time from a biopsychosocial perspective, leading to a more holistic approach to pain and comfort management.

The Institute of Medicine's (IOM) report, *Relieving Pain in America* (2011), identifies that pain management is more than just the administering of pain medication. Successful pain management is largely dependent upon the strength of the relationship between the clinician, patient, and patient's family (IOM, 2011). The Nichols–Nelson Model of Comfort (NNMC) is consistent with the recommendations made by the IOM, that pain management must go beyond pain medication to include relationships. The NNMC includes seven dimensions of comfort as well as predictors of comfort that are both internal and external to the patient.

Successful pain management is largely dependent upon the strength of the relationship between the clinician, patient, and patient's family (IOM, 2011).

The Shift: The Physiology of Pain Includes the "Physiology of Comfort"

Competent management of both pain and comfort requires an understanding of the mind–body response to pain. The physiology of pain includes four components: transduction, transmission, modulation, and perception.

1) *Transduction* is the physiological process by which a stimulus (mechanical, chemical, or thermal) is converted into an electrical impulse. During transduction, nociceptors (sensory receptors) are activated when they are stimulated by such things as pressure, heat, or chemical irritation (Berry et al., 2001; Pasero & McCaffery, 2010).
2) *Transmission* is the movement of the electrical impulse from the point of transduction to the spinal cord along nerve fibers. The final stage of transmission is termination in the brainstem and thalamus to notify the patient that there is pain (Berry et al., 2001; Pasero & McCaffery, 2010). The processes of transduction and transmission continue as long as the stimulus is present.
3) *Modulation* is the inhibition of the transmission of messages of pain to the brain by using the body's release of hormones such as serotonin, norepinephrine, and endogenous opioids (Pasero & McCaffery, 2010).
4) *Perception* is the brain's interpretation of the pain quality, including location and intensity of the pain (Pasero & McCaffery, 2010).

Nichols asserts that the point at which comfort gives way to pain or pain gives way to comfort is within the third and fourth components: modulation and perception (2018). This assertion is supported by research in the human response to pain and the pursuit of comfort, which has identified that the perception of pain is influenced by the affective-emotional states of the person experiencing the pain (Berry et al., 2001; Pasero & McCaffery, 2010), as well as emotional states such as fear and anxiety (Backonja et al., 2010). Emotional states in the process of pain perception are processed in the limbic system and modulate the amount of pain a person perceives (Hansen & Streltzer, 2005). The signal within the limbic system interacts with information from the environment as well as past memories, which then influence the perception of pain (Hansen & Streltzer, 2005). Other emotional responses to pain arise from the anterior cingulate and the right ventral prefrontal cortex, which are activated by social rejection (Hansen & Streltzer, 2005). Bayet, Bushnell, and Schweinhardt (2014) found that just viewing negative facial images modulates perceived pain intensity and unpleasantness. Rainville, Bao, and Cretien (2005) found that negative emotions and the desire for relief influence pain perception.

These physiological responses to pain provide a substantive argument for a theory positing that the connection between pain and comfort lies between the third and fourth components—modulation and perception—of the physiology of pain/comfort. Considering the power of modulation and perception to provide or prevent the experience of comfort, we see the degree to which context can impact pain perception and the subsequent pain and comfort experience. Context includes the environment, past and present memories, thoughts, and, most importantly, events occurring in the moment. The NNMC proposes that this expansion of pain physiology to include the physiology of comfort, sets us up to attend to a wider variety of clinical conditions, and is therefore, a more appropriate theory on which to base future studies on pain relief.

Further, Nichols theorizes that allowing the individual to provide a narrative is essential to adequately understanding the individual context and experience of the person's pain (2017, 2018). When clinicians carefully listen to the individual's story, which includes the context of pain, it helps the individual feel believed and thus enhances patient–clinician trust.

When clinicians carefully listen to the individual's story, which includes the context of pain, it helps the individual feel believed and thus enhances patient–clinician trust.

Clinicians' Beliefs Drive Their Practice

Considering the significant impact of pain on patients, healthcare professionals must ask themselves what they believe regarding pain and what they believe their role is in helping the patient find comfort. These beliefs are created, in part, through the processing of information that relates to clinical care and the professional role. The information clinicians are processing, however, may be complete, incomplete, accurate, or inaccurate. Either way, their processing of this information shapes their beliefs. Clinicians learn in school that pain is a natural and important human process. There are currently several biomedical theories of pain, as noted in Table 13.1.

These biomedical theories of pain, which have guided clinicians since 1955, do not consider that the patient's pain could

Table 13.1   Biomedical theories of pain.

Theory	Summary statement
Pattern theory (Sinclair, 1955)	Proposes that pain's nerve impulse pattern is produced by intense stimulation of nonspecific receptors because the body has no specific pain fibers or endings.
Specificity theory (Sweet, 1959)	Proposes that pain is a specific modality, like vision or hearing, with its own central and peripheral apparatus.
Descarte's model of pain (Melzack & Wall, 1965)	Proposed a linear mechanism of pain transmission, from the periphery through the spinal cord to the brain.
Gate control theory of pain (Melzack & Wall, 1965)	Proposed that pain is a function of many parts of the nervous system, not just one specific part. Presents the concept of impulses reaching and exceeding a critical level. Provided the explanation of the nervous system and spinal cord regulating pain as a gating system inhibiting or facilitating a noxious stimulus. The theory acknowledges that biological, cognitive, behavioral, and affective elements interact to determine whether a stimulus is inhibited or facilitated. No psychosocial influence is mentioned.
Neuromatrix of pain (Melzack, 2001)	Expands the Gate Control theory to include a genetically predetermined widely distributed neural network that is modified by sensory experience. Makes no mention of a specific mechanism of action, but it is proposed that the neuromatrix is impacted by injury, pathology, and chronic stress leading to activation of homeostatic and behavioral programs.

be increased or decreased by the actions, inactions, or way of being of the care provider or the healthcare system. As you can imagine, patient survey items derived from biomedical theories do not leave any room for the patients' own stories of their pain. Exclusion of potentiating factors from the assessment of pain can lead to an incomplete plan of care and subsequent unresolved discomfort, loss of hope, and prolonged or persistent pain. Comfort is a dynamic construct that requires clinician and patient to examine the context, culture, and beliefs of both the patient and the care provider. When clinicians do not understand their role in pain and comfort management, their disposition can communicate disbelief to the individual in pain. This perception will impact the patient–clinician relationship and the plan of care. The patient–clinician relationship plays an essential role in the development and refinement of the plan for additional comfort modulators. A well-developed plan of comfort does not exclude the possibility of the patient having pain,

When clinicians do not understand their role in pain and comfort management, their disposition can communicate disbelief to the individual in pain.

but will instead help the patient cope in a different way with the presence of pain.

To help clinicians and patients make this shift from pain management to comfort management, it is important to understand the difference between these two ways of thinking. Appendix N explains the differences between a focus on pain and a focus on comfort.

Assumptions of the NNMC based on the patient–clinician relationship include the following:

- Therapeutic use of self by healthcare professionals increases comfort,
- Comfort occurs through biogenic caring[1] in the lived patient experience, and
- Comfort and caring are connected (Nichols, 2018).

Clinicians can provide comfort only if they are caring.

If these assumptions are correct, clinicians can provide comfort only if they are caring (Nichols, 2018). The theory behind the NNMC is that human connection within a therapeutic patient–clinician relationship is what drives the modulation of the dimensions of comfort. According to this theory, the quality of relationships must be considered in addition to the patient's report of pain and comfort if the process of comfort management is to be successful.

Dimensions of Comfort

Nichols (2018) describes the seven dimensions of comfort as mental states experienced by the person in pain. The seven dimensions, which will be explained fully in the next few pages, are (a) fear, (b) pain, (c) suffering, and (d) grief/loss of function, all of which the clinician seeks to decrease, and (e) trust, (f) well-being, and (g) perceived inclusion, all of which the clinician seeks to increase. Nichols proposes that if a strong patient–clinician connection exists, the clinician can identify which of these mental states is modulating the individual toward or away from comfort.

---

1 Biogenic caring is defined as caring that is life-giving and life-receiving for both parties (watsoncaringscience.org).

All of the dimensions of comfort are either present or absent to varying degrees, and are therefore best seen as existing on a continuum. In the context of a formal study, in order for a continuum to be informative, the extremes must fit reasonably on opposite ends of the continuum. Therefore, if we were to put together a study related to comfort, we would want to look at predictors that tested as opposites. For example, if we are measuring internal predictors, it is important to ask the patient if they have fear, and then ask them in a separate item if they have no fear. If these are the opposites, patients who strongly agree with an item that states, "I have fear," will also strongly disagree with an item that states, "I have no fear."

Dimensions of Comfort That Clinicians Seek to Decrease

Fear

The ideal state related to the comfort dimension of fear is relief and/or the subsiding of distressful emotions aroused by the unknown, the unclear, or harmful and painful experiences, real or imagined. *Desired patient state-of-being:* "Now that I believe you see me, I feel held and protected; I am at ease."

To measure fear as perceived by the patient, a statement about fear will be presented to the patient to understand where on the continuum of fear the patient is. Fear, like all seven dimensions of comfort, can range from absent to present. To measure the degree of fear perceived during the experience of pain, a statement about fear is presented to the patient, using a 7-point Likert scale. The statements for fear are "I am experiencing fear" and "I am not experiencing fear," and the patient responds to these statements, selecting anywhere from strongly disagree (1) to strongly agree (7).

Measuring the patient's degree of fear, as with all the other dimensions of comfort, will aid in action planning for comfort. Equally important, measurement will enable further study of fear as a dimension of comfort. Using factor analysis, we can discover how fear holds together with other dimensions of comfort and whether it is a leading factor in the construct of comfort—or whether it belongs at all. Measuring each dimension of comfort individually enables us to study the construct of comfort scientifically. Studying how dimensions like fear fit, informs the theory itself, so the theory can be refined and then retested as an updated theory of comfort.

Pain

The ideal state related to the comfort dimension of pain is diminished and/or altered perception of physical suffering and discomfort. *Desired patient state-of-being:* "Now that I believe you see me, I am in a state of comfort; I am at ease and can focus my energy to cope with this state of vulnerability."

Patient survey items related to this dimension of comfort include "I have no pain" and "I am experiencing pain."

Suffering

The ideal state related to the comfort dimension of suffering is diminished perception of the negative psychological and/or physiological impact of physical pain, disability, loss, fear, anxiety, confusion, and frustration surrounding the experience of being vulnerable (Dempsey, 2014). *Desired patient state-of-being:* "Now that I believe you are my advocate, I choose to partner with you in focusing my being and my energy toward healing."

Patient survey items related to this dimension of comfort include "I am not suffering" and "I am suffering."

Grief/Loss of Function

The ideal state related to the comfort dimension of grief/loss of function is experiencing external and internal acceptance and understanding as it relates to loss of function. *Desired patient state-of-being:* "Now that I believe you see me, I will not be defined by my loss but can begin to accept the change it brings."

Patient survey items related to this dimension of comfort include "I have grief related to loss of physical function that occurred because of my pain" and "I have accepted the loss of physical function that occurred because of my pain." (Note: the patient would have been asked prior to this whether they have experienced a loss of physical functioning because of their pain, answering, Y/N.)

Dimensions of Comfort That Clinicians Seek to Increase

Trust

The ideal state related to the comfort dimension of trust is renewed confidence in a person or institution on which one relies. *Desired patient state-of-being:* "Now that I believe you see me—and I know that you are genuinely interested in me, my values—and my beliefs, our relationship is renewed."

Patient survey items related to this dimension of comfort include "I trust in the people caring for me" and "I have no trust in the people caring for me."

Well-being

The ideal state related to the comfort dimension of well-being is experiencing a positive interaction between one's circumstances, activities, and psychological resources. *Desired patient state-of-being:* "Now that I believe you see me, I am willing to participate in the plan of care and take ownership of the outcomes."

Patient survey items related to this dimension of comfort include "I have a sense of well-being" and "I do not have a sense of well-being."

Perceived Inclusion

The ideal state related to the comfort dimension of perceived inclusion is feeling connected to care and treatment while in a vulnerable state. *Desired patient state-of-being:* "Now that I believe you see me, I am willing to listen to the knowledge you possess to guide my journey of healing. We are connected in a shared vision for my health, and I can rest while I heal."

Patient survey items related to this dimension of comfort include "My care team here includes me in planning my care" and "My care team here excludes me in planning my care."

An NNMC tool exists to study all of these variables and more, related to the patients' experience of pain and comfort. The tool to measure the patient experience of comfort is in Appendix O, and the tool to measure the provider experience is in Appendix P.

Predictors of Comfort

Both external and internal predictors of comfort can modulate a person's perception of comfort and pain.

External Predictors of Comfort

External predictors of comfort are the clinician's "energy" and behaviors toward the patient in the patient's lived pain experience. Specific external predictors include the clinician's knowledge, skills, and attitudes. External predictors also include staff members':

1) Beliefs about comfort;
2) Clarity about themselves, their role, and the care delivery system;
3) Therapeutic competence;
4) Skill in interprofessional collaboration; and
5) Level of professional development.

An instrument that measures all of these variables and more, related to the clinicians' experience, is provided in Appendix P. Measurement of each of these external predictors will enable clinicians to make more informed choices about how they will work to manage pain and comfort, and perhaps just as importantly, how they can improve their own "therapeutic use of self" (Taylor, 2008) as they tend to the pain and comfort needs of their patients.

Nichols proposes that external predictors can also intensify internal predictors. Thus, internal predictors can be direct modulators of comfort or indirect modulators of comfort, secondary to the impact of external predictors.

Internal Predictors of Comfort

Internal predictors of comfort are the beliefs and thoughts of the person in pain. An extreme form of this would be pain catastrophizing where a patient ruminates on their pain and believes they will never find relief, which leads to feeling hopeless and helpless (Sullivan, Bishop, & Pivik, 1995). During modulation and perception, internal predictors can either enhance the positive dimensions of comfort clinicians want to increase or inadvertently increase the negative dimensions of comfort clinicians wish to decrease (Ossipov, Morinura, & Porreca, 2014). Only if the internal predicators improve the mindset of the person in pain will comfort be enhanced. Specific internal predictors in the NNMC that intensify pain and detract from comfort include worry, anger, bitterness, helplessness, hopelessness, loneliness, and feeling valueless. Patient survey items related to each of these conditions are contained in the NNMC tool to measure the patient experience of comfort, found in Appendix O.

The Model

The Nichols–Nelson Model of Comfort (NNMC) is a theoretical model that depicts the predictors and outcomes of comfort. Figure 13.1 provides a review of the model.

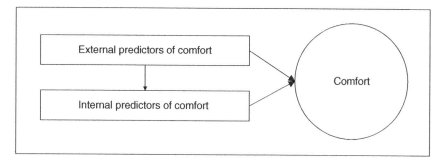

Figure 13.1   Nichols–Nelson Model of Comfort (NNMC).

The authors of this chapter believe the NNMC has implications to:

1) Guide practice,
2) Direct translational research (research that can be applied to operations),
3) Improve the patient experience, and
4) Guide future research on the patient–clinician relationship and the lived pain experience.

Summary

In the work of predictive analytics, theory is essential for:

1) Guiding the conversation about the hunch (what you believe causes the outcome),
2) Developing measurement instruments based on that hunch, and
3) Interpreting results from the psychometrically tested instruments.

The optimization of any quality improvement project using predictive analytics requires beginning with the best possible theory.

The optimization of any quality improvement project using predictive analytics requires beginning with the best possible theory. As the work of continually improving care moves forward, certain outcomes will be targeted for improvement. If one of those outcomes is related in any way to pain control, working with a theory such as the NNMC will help your efforts be more realistic, effective, and patient-focused.

If your efforts are focused on any other outcome, it is essential to carefully examine and fully understand the quality and appropriateness of the theories you are using. Even theories that are highly relevant can be focusing on only one aspect of a multifaceted story.

14

Theory and Model Development to Improve Recovery from Opioid Use Disorder

Alicia House, Kary Gillenwaters, Tara Nichols, Rebecca Smith, and John W. Nelson

Many clinicians in the field of addiction recovery have long held the perspective that healthy relationships—specifically a client's healthy relationship with at least one trusted, caring other—are central to the person's ability to recover from opioid use disorder (OUD), which is a subset of substance use disorder (SUD). Research also supports this perspective (Fallin-Bennett, Elswich, & Ashford, 2020; Jack, Oller, Kelly, Magidson, & Wakeman, 2018); and yet, measurement of the presence or absence of a therapeutic relationship with a trusted other is missing from current models designed to assess the effectiveness of recovery from OUD. Until OUD places healthy, trusting relationships in the center of treatment, outcomes related to OUD recovery will continue to tax healthcare systems and devastate the human family. This chapter reviews how the presence of healthy relationships, along with measurement of the effectiveness of those relationships, will improve the science, operations, and outcomes in OUD recovery.

The Current Costs of Opioid Use Disorder (OUD)

Opioid use disorder (OUD) remains a significant health concern in the United States. In 2016 there were over 11.5 million adults who reported misuse of prescription opioids and almost 2 million who reported an opioid use disorder (Christie et al., 2017). Estimates of the annual financial cost of opioid abuse in the

Using Predictive Analytics to Improve Healthcare Outcomes, First Edition.
Edited by John W. Nelson, Jayne Felgen, and Mary Ann Hozak.
© 2021 John Wiley & Sons, Inc. Published 2021 by John Wiley & Sons, Inc.

United States range from $80 to $500 billion (Dunlap et al., 2018). The incidence of death from opioid abuse continues to increase despite numerous interventions and abundant funding to fight the problem (Itzoe & Guarnieri, 2017).

Interventions for OUD

The most common intervention to promote recovery from OUD is not the most successful intervention.

The most common intervention to promote recovery from OUD is *not* the most successful intervention. The most common recovery interventions used in healthcare settings are cognitive/behavioral therapies, which are used with up to 59% of patients (Wakeman et al., 2020). By contrast, the most effective intervention to promote recovery from OUD, by research standards, is medication-assisted therapy (MAT); yet it is used with only 2.4%–12.4% of patients (Wakeman et al., 2020). The primary medications used for MAT are methadone, an opioid agonist; buprenorphine, an opioid partial agonist; and naltrexone, an opioid antagonist.

It is a realistic concern that people and organizations working with individuals in recovery, particularly in bottom-line-driven environments, may look at the high success rates of medication-assisted therapy and choose MAT over cognitive/behavioral health therapy, appreciating, in part, that the improved outcomes seem to be happening with less human contact—or in administrative terms, with fewer full-time employees (FTEs). However, there is strong evidence that MAT is especially effective when used in combination with cognitive/behavioral therapies (Rezapour et al., 2019; Sofuoglu, DeVito, & Carroll, 2019). Despite what might be assumed from a quick look at the difference in recovery outcomes between patients receiving cognitive/behavioral therapy alone and those receiving MAT alone, the positive effects of cognitive/behavioral therapy (also known as cognitive or behavioral training) should be built upon rather than discontinued.

Introduction of a "Trusted Other"

Significantly, cognitive/behavioral therapy is also one of the ways a recovery program can provide the person in recovery with a "trusted other," the value of which should not be underestimated. People in recovery are never simply stopping a behavior. They are simultaneously re-evaluating their entire lives. They are exploring what it was in their lives and histories that caused them to misuse opioids. They may also be eliminating or reshaping unhealthy relationships and forming and nurturing new relationships. Medications such as methadone, buprenorphine, and naltrexone may help put the person in recovery in a better frame of mind to re-evaluate their life, but they can obviously provide none of the essential guidance needed for such a multilayered undertaking.

It is possible, and we believe it is worth testing, that a healthy, stable relationship with a trusted other who causes people in recovery to feel "able" will be a key predictor of successful long-term recovery. Much like the function of a sponsor in Alcoholics Anonymous who is there for the person to call on, not only in times of weakness and struggle but throughout a process of reassessing the person's life, the presence of a healthy relationship with a trusted other provides the person with needed guidance *and* a reliable, steady, sturdy person on whom to lean.

To understand whether the presence of a trusted other or relationships more generally have been examined in relationship to OUD, the literature was searched for predictive analytics studies that included relationships in the study of recovery from OUD.

In total, only three studies were found that used predictive analytics to study opioid use disorder (Ciesielski et al., 2016; Cochran et al., 2014; Zamirinejad, Hojjat, Moslem, Moghaddam Hosseini, & Akaberi, 2018). Ciesielski et al. (2016) examined risk for OUD in a sample of 694,851 patients who were initially prescribed opioids for pain management without previous diagnosis of OUD. Among the 2,064 patients who developed OUD, both patient and demographic variables were found to predict OUD (Ciesielski et al., 2016). It is proposed that the 14 variables identified could be used to screen for risk (Ciesielski et al., 2016). Cochran et al. (2014) studied 2,841,793 opioid prescription claims and found eight variables that predicted OUD for 2,913 of the patients within the database. Zamirinejad (2018) studied maladaptive behaviors as predictors for developing OUD in a study of 120 patients diagnosed with OUD with 60 men who tested positive for maladaptive behaviors and 60 men who did not. Maladaptive behaviors were found to be a predictor of onset of OUD at an earlier age, and Zamirinejad (2018) proposed that a history of maladaptive behaviors should be considered when assessing patients at risk of developing

OUD. A crosswalk of the variables from all three studies revealed that the only risk factor reported was "psychiatric issues" (Ciesielski et al., 2016; Cochran et al., 2014; Zamirinejad et al., 2018). Two of the studies identified the additional factors of being male, younger age, and having a history of using multiple pharmacies in an attempt to get more medications (Ciesielski et al., 2016; Cochran et al., 2014). All risk factors from all the studies are noted in Appendix Q.

Examination of the literature regarding the use of predictive analytics in the treatment of OUD revealed no research that pursued a trajectory that includes relationships.[1] We believe the examination of the presence of a trusted other in the life of a person attempting to recover from OUD will be a significant predictor of long-term recovery.

Since there is already evidence that MAT and cognitive therapy (especially in combination) are effective, but rates of sustainable recovery from OUD are still dismally low, we propose the addition of the variable "trusted other" to the model for sustained recovery from OUD as noted in Figure 14.1. Later in this chapter, we will discuss how to measure this latent variable. Figure 14.1 depicts how this belief would be measured.

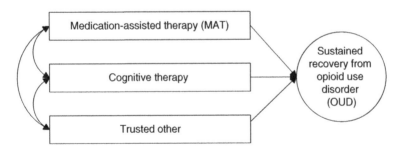

Figure 14.1 A proposed model of OUD treatment, which includes the variable of trusted other.

1 Nine databases were examined, with the search limited to studies that were (a) reported in English, and (b) based on data collected in the previous 5 years. Databases included Web of Science, Biological Abstract, BIOSIS Citation Index, CAB Abstracts, Current Contents Connect, Data Citation Index, Derwent Innovations Index, MEDLINE, and SciELO Citation Index.

Pain Management, OUD, and Therapeutic Relationships

Approximately 62% of opioid abuse is reported by those attempting to control physical pain (CDC, 2017). In many situations, acute pain can appropriately be treated with opioids, but including opioids in the treatment of people with chronic pain (whether they are experiencing OUD or not) is controversial (Donroe, Holt, & Tetrault, 2016; Langford, 2017; Mehta & Langford, 2006). Nearly all healthcare clinicians, regardless of specialty, will care for patients with acute and/or chronic pain at some point (Donroe et al., 2016; Olsen, 2016), and we know that many healthcare professionals are reluctant to treat people with opioids (Donroe et al., 2016; Langford, 2017; Olsen, 2016). That reluctance can put stress on the clinician–patient relationship. This fact is troubling because relapse of OUD is more often triggered by inadequate pain treatment than by the short-term use of opioids (Donroe et al., 2016; SAMHSA, 2020). Conversely, some people with OUD have worked hard to get off opioids, and they are willing to try anything to treat their pain before taking another opioid. Most distressing are the complaints of patients with chronic pain of the abrupt discontinuation of opioids, which can lead to illicit drug use (Donroe et al., 2016; SAMHSA, 2020).

Relapse of OUD is more often triggered by inadequate pain treatment than by the short-term use of opioids.

In addition to stressed clinician–patient relationships, we must consider the potential for additional trauma from "biopsychosocial" pain (Nichols, 2018, 2019). Biopsychosocial pain can develop in patients due to poorly treated pain, clinicians not believing patients are in pain, and patients feeling like they are being judged or stigmatized as drug addicts. Biopsychosocial pain is further exacerbated when patients in pain experience "loss of hope" and increased isolation due to increased avoidance by healthcare professionals, friends, family, and social groups (Langford, 2017; Nichols, 2018, 2019; Olsen, 2016). Thus, central to building trust is a therapeutic relationship with a trusted other (Nichols, 2018, 2019).

This myriad of challenges in managing pain warrants having a trusted other to talk through these issues and together find solutions to achieve and sustain recovery. In many clinical education and certification programs, the development of technical skills is foundational, but very little is taught or required for the development of therapeutic or relational skills (Koloroutis

& Abelson, 2017; Koloroutis & Trout, 2012). In order to develop a trusting, therapeutic, supportive relationship with any patient, relational skills are essential, and we would submit that these skills are even more essential for clinicians working with patients with chronic pain.

Interventions Which Include Potential Trusted Others

Several programs and interventions shown to improve outcomes related to opioid use are relationship-based, thereby providing patients with clinicians and others who may eventually function as their trusted others. For example, Narcotics Anonymous, which helps participants connect or reconnect relationally to a higher power and also includes a fellowship component, has been shown to help sustain remission from opioid use for 1 year (Shiraly & Taghya, 2018) and, in a later study, for up to 42 months (Weiss et al., 2019). The study by Shiraly and Taghya (2018) revealed that family support was also an important variable for sustaining remission from opioid use. Counseling has been found to sustain abstinence from opioid use in the postpartum period for new mothers (Martinez & Allen, 2020). Counseling in how to cope with chronic pain was found to decrease opioid use for patients who reported addiction to opioids to manage chronic pain (Messina & Worley, 2019). Significantly, all of these interventions are relationship-based; they relate to the person's relationship with self, with a higher power, and with supportive people in the person's life, including caring clinicians.

Common Relationships in OUD Recovery Programs

Various relationship-based roles and structures within OUD recovery care are found in the literature to be effective, including: (a) case management, (b) recovery/connection coaches, (c) recovery management checkups, (d) peer case managers/peer outreach workers, and (e) professional addiction counselors.

Case Managers

According to the Treatment Improvement Protocols (TIPs) by the Substance Abuse and Mental Health Services Administration

(SAMHSA), case management, in the setting of substance abuse treatment, is an integral component for improved treatment and recovery outcomes (SAMHSA, 2020). Additionally, substance use disorders (SUDs) tend to be comorbid with other mental health issues or disorders which necessitates comprehensive management and service coordination for patients and clients needing various levels of care. Case managers are present throughout the care continuum from case finding and pretreatment, to primary treatment, and finally to aftercare.

Recovery/Connection Coach

Similar to case management is a role called the recovery/connection coach, which exists to meet individual client needs and provide the connection to services deemed necessary to sustain abstinence (Jack et al., 2018). These are professionals who have had their own addiction experiences and can relate to the patient recovering from a substance use disorder (Jack et al., 2018). This role is increasingly used and is shown to help patients navigate the system and connect to resources (Jack et al., 2018). Peer recovery coaches are being integrated into rehabilitation programs, detox clinics, and other community health settings to engage, support, and promote treatment among individuals with OUD (Sightes et al., 2017).

Recovery Management Checkups

To address the chronic nature of most SUDs—characterized by cycles of relapse, treatment reentry, and recovery—the use of recovery management checkups (RMCs) has emerged as a potential solution for post-discharge monitoring and/or aftercare (Scott, Dennis, & Foss, 2005). Individuals whose aftercare included RMCs were significantly more likely to voluntarily return to treatment sooner and receive more treatment (Scott et al., 2005). In order for this structure to provide a person in recovery with a trusted other, care would have to be taken to provide continuity of relationships for recovery managers and their clients.

Peer Case Managers, Outreach Workers, and Recovery Specialists

Improved treatment engagement and client satisfaction have been achieved with the assistance provided by peer case managers, especially for homeless OUD clients (SAMHSA, 2020). Peer outreach workers have also been utilized to identify out-of-treatment individuals with OUD to increase the provision and sustainability of medication-assisted therapy (MAT) and to minimize opioid-related overdoses and related fatalities (Scott, Grella, Nicholson, & Dennis, 2018). Most of the published studies on peer support programs, peer providers, and their role in SUD recovery, indicate positive outcomes, including decreased rates of hospitalization and increased medication adherence; however, the research regarding peer support remains inconclusive (Blash, Chan, & Chapman, 2015).

Addiction Counseling Professionals

Addiction counselors may be court-appointed or engaged privately by clients in any stage of recovery. Counseling professionals use a variety of approaches from behavioral to root-cause approaches. Contact between clients and counseling professionals can be one-to-one, in group settings, or a combination of the two. Effectiveness of these relationships is improved when clients feel the counselor cares about them as a person, believes in their ability to advance, and will remain a sturdy presence for them and *with* them, over time (Joe, Knight, Becan, & Flynn, 2014; Olmstead, Abraham, & Martino, 2012). This relationship is critical during both the inpatient treatment portion of OUD recovery and in longer term abstinence after inpatient treatment (Joe et al., 2014). The first week of treatment has been found to be critical in the establishment of a relationship with the person in recovery (Joe et al., 2014).

Because the addition of each of these relationship-based roles and structures has led to successful outcomes, it is proposed that any or all of them could be productively integrated into a relationship-based predictive model in OUD recovery. If any of them are currently part of your day-to-day operations, or if you are considering adding a new role or structure, its effectiveness should be measured, ideally from the outset. While a person in any of these roles could theoretically become the "trusted other" who helps with long-term recovery, we are also keenly aware that a person in each of these roles must have a high level of self-awareness and overall relational competence to be effective in these roles. Any assessment of the effectiveness of these roles should not be seen as more important than assessments of the effectiveness of the individuals enacting each of these roles.

Existing OUD Measurement Instruments

Instruments currently used to assess OUD are premised on the American Society of Addiction Medicine (ASAM) criteria (Mee-Lee, 2013). The criteria were developed through a collaborative effort that began in the 1980s and is the most comprehensive set of guidelines used widely in the United States. In fact, their use is mandated for addiction services in 30 states. The multidimensional assessment using the ASAM criteria contains six dimensions which are deemed important for recovery from substance abuse: (a) acute intoxication and/or withdrawal potential; (b) biomedical conditions and complications; (c) emotional, behavioral, or cognitive conditions and complications; (d) readiness to change, (e) relapse, continued use, or continued problem potential; and (f) recovery/living environment. Along with these six dimensions, the addition of organization-specific factors can facilitate more comprehensive and context-specific analysis of outcomes.

Given that the ASAM criteria are so broadly used and are embedded within a great deal of the work of addiction medicine, it made sense to integrate them into our initial model which included the variables of trusted other, cognitive therapy, and MAT. An updated model to be used in predictive analytics to study OUD recovery which integrates the ASAM dimensions with our initial three variables is provided in Figure 14.2

Updating the Old OUD Measurement Instrument and Model to Include the Trusted Other

The challenge with an updated model like the one in Figure 14.2 is that the latent variable of "trusted other" requires measurement of a variable that is experiential and not directly observed. Thus, we propose measuring those elements of the trusted other which we believe assist with progression toward substance recovery. As the model in Figure 14.1 indicates, we believe that

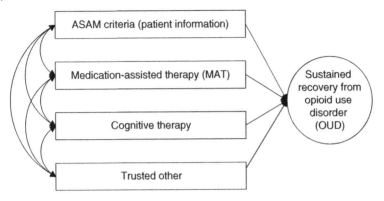

Figure 14.2 A model of OUD treatment integrating ASAM criteria, treatment modalities, and trusted other.

MAT, cognitive therapy, and the previously missing variable of the trusted other with whom the person in recovery can have a stable, healthy relationship, are a promising formula for long-term recovery. The following survey items would provide insight into the latent variable of the trusted other.

We would begin with instructions to help patients and clients identify their trusted other(s):

For the purposes of the following eight questions, your "trusted other" is a person you trust, who you are in contact with during recovery, and who you are able to speak with on a regular basis. This could be a professional clinician, coach, or peer who is provided by your recovery program, or it could be someone with whom you have a purely personal relationship. In short, your trusted other is anyone you trust and feel you can consistently confide in.

Survey items:

1) To what extent does your trusted other believe in your ability to stay opioid free?
2) To what extent do you believe in your ability to stay opioid free?
3) To what extent does your trusted other believe that your medication will help you stay opioid free?
4) To what extent do you believe that your medication will help you stay opioid free?
5) To what extent does your trusted other believe in your ability to make some improvements in your life?

6) To what extent do you believe in your ability to make some improvements in your life?
7) To what extent does your trusted other believe in your ability to make long-term, sustainable improvements in your life?
8) To what extent do you believe in your ability to make long-term, sustainable improvements in your life?

The client would then be asked to list who this person is either by name or role or relationship. It may be that the client identifies more than one trusted other, in which case a survey would be given for each of the relationships identified.

We hypothesize that a trusted other's belief in the client leads to clients believing in themselves, which predicts clients feeling "able." This ability can be explained as a sense of personal agency—simply a movement from "I don't believe I can" to "I believe I can." This sense of "I believe I can" aids in every stage of recovery, from the initial decision to stop using, to a series of daily decisions not to use, and throughout long-term recovery. The relationship with trusted others is particularly critical at a time when people are experiencing post-acute withdrawal syndrome for the first six months to two years in recovery (Gorski & Miller, 1986; Hazelden Betty Ford Foundation, 2020). The trusted other is an essential copilot at a time when the healing of the nervous system includes periods of foggy thinking, difficulty with problem solving, irritability, etc. (Gorski & Miller, 1986; Hazelden Betty Ford Foundation, 2020).

A quantification of the unseen but powerful variable of the trusted other would allow evaluation of this relationship empirically as it relates to the outcome of recovery from opioid addiction. In our initial proposal, the assessment would measure both the client's perception of the clinician's belief and how much clients believe in their own abilities to grow, change, and recover.

Discussion

Given that we do not know which of the roles described in this chapter is most effective as a vehicle to provide a trusted other for patients, it is suggested that organizations intending to adopt a model of care that includes a "trusted other" begin by assessing what they have in place to identify (a) the abilities of patients to trust others, (b) the level of training of clinicians to enact each role, (c) the relational competence of clinicians in these roles, and (d) the structures already in place to support the relationship between patients and clinicians. Along with structures to support relationships between patients and clinicians, it is important to form structures that foster high competence in interprofessional and interdisciplinary practice. We may find that the quality of relationships among healthcare professionals correlates with long-term OUD and substance use disorder (SUD) recovery. Do organizations that demonstrate more collaborative relationships among a variety of clinicians, practitioners, and healers yield higher rates and durations of recovery? Does increased understanding of interoception (understanding one's own internal sensations), by both patient and clinicians correlate with increased recovery rates? When patients and families rate the respect across disciplines and/or for patients and families higher, do recovery rates improve?

As you can see, we have a great deal to learn about what it will take to provide a trusted other for people recovering from OUD.

Broader Recommendations for People Designing Recovery Programs

An essential question in this work is, "What qualities help turn clinicians into trusted others?" People seeking treatment for OUD or any other SUDs are not a monolithic group, so devising a tidy checklist of ideal qualities for clinicians would not be useful. For example, some people in treatment will respond to a "tough love" approach and others will respond to a more understanding, empathetic approach. That said, we do believe there are some personal qualities and therapeutic practices that will help clinicians become the kind of trusted others who can be of effective long-term service to people in recovery. They are (a) modeling "not knowing," (b) cultivating self-awareness, and (c) committing to seeking the perspectives of others. While we have sought to delineate these three qualities from each other, we have found, as we worked with them, that they are essentially all the same thing: people with high self-awareness tend to be keenly aware of what they do not know (and quite comfortable with the sensation of not knowing), and as a result, they are good at seeking and considering the perspectives of others.

A comprehensive exploration of these three qualities and practices is found in Appendix R.

Advancing a Culture That Supports Clinicians to Become Trusted Others

If an organization's aim were to design a system that would nurture clinicians to become trusted others, the organization would strive to embrace the same characteristics it wants to foster in its clinicians—modeling not knowing, cultivating self-awareness, and adopting an attitude of perspective seeking—but at the system level. If you are involved in any way in the recovery industry or a recovery sector of a healthcare system, it is essential that you get clarity about exactly what "business" you are in. Are you and your organization in the business of helping people stop using addictive substances, or are you and your organization in the business of helping people with SUDs to stop using, rebuild their lives, and thrive into the future? A good place to start this inquiry is with any existing mission or vision statements. But once you have either of those statements in hand, it is important to make a study of the degree to which the structures, processes, and policies in your organization support your stated mission or vision. It is not enough to claim to be in the thriving business and then provide little or nothing that is directly in service of helping people thrive.

For an exploration of the ways organizations and systems can foster environments better equipped to address OUD cessation, see Appendix S.

Conclusion

We submit that people in recovery from OUD and other SUDs are in need of the exact three things that we have prescribed for clinicians who wish to be trusted others and organizations that wish to gain the ongoing trust of their clients: the ability to tolerate not knowing, a life-long commitment to self-awareness, and an attitude of perspective seeking. In the ideal recovery situation, a patient would work with a trusted other who is personally fostering these three qualities within a system that is fostering them on a systemic level. As a person works to recover from a potentially life-ruining (or life-ending) addiction, imagine the comfort that would come with the realization that it is possible to cultivate these qualities in one's self. The patient is our reason for being. Our aim should be to help people recover from OUD and other SUDs and to learn what it takes for each patient, individually, to thrive.

Much of what has been proposed here is yet to be researched using a measurement model such as the model noted in Figure 14.2. Teams anywhere wishing to test this new model are enthusiastically invited to do so.

Section Four

International Models to Study Constructs Globally

15

Launching an International Trajectory of Research in Nurse Job Satisfaction, Starting in Jamaica

John W. Nelson and Pauline Anderson-Johnson

Conducting research can be exciting, messy, disappointing, and illuminating. It is important to understand how to navigate both the successes and challenges in research in order to successfully apply your findings in operations and to disseminate your study for replication. For the study discussed in this chapter, a research model was developed based on a hunch that there may be a relationship between nurses' clarity of self, role, and system, and nurse job satisfaction. The model was developed using theory and measurement instruments conceived in the United States, but there was a strong desire to study this model internationally. We began our international work in Jamaica to clarify, worldwide, what contributes to nurse job satisfaction. This chapter reviews how building a body of work internationally requires navigation of cultures and adaptation of measurement instruments to accommodate cultural nuances in language and context. It will also include how the challenge of a small response rate was managed using parceling of data, as well as an interpretation of curious findings which related to the Jamaican culture.

Background

This study sought to examine (a) the validity and reliability of a US-tested multifaceted instrument to measure nurse job satisfaction when used in Jamaica, and (b) whether clarity of

Using Predictive Analytics to Improve Healthcare Outcomes, First Edition.
Edited by John W. Nelson, Jayne Felgen, and Mary Ann Hozak.
© 2021 John Wiley & Sons, Inc. Published 2021 by John Wiley & Sons, Inc.

self, clarity of role, and/or clarity of system predicted nurse job satisfaction in Jamaica. This study was the beginning of a larger international research trajectory that sought to test nursing-specific measurement instruments and concepts conceived in the United States. It was also a pilot study to help make decisions about additional international research.

Pilot studies can inform researchers about the ways in which variables relate to one another, and the findings from those studies can be used in further study (Kistin & Silverstein, 2015; Leon, Davis, & Kraemer, 2011). Pilot studies are particularly valuable when the research is novel (Leon et al., 2011). Though pilot studies are rarely published, we believe they should be because of what they can teach us about how to more efficiently and effectively conduct a larger study on the same subject (Lancaster, 2015). Unfortunately, only 8.8% of pilot studies actually lead to larger studies (Arain, Campbell, Cooper, & Lancaster, 2010). Adequate pilot studies require 20–50 participants per study group (Hertzog, 2008; Johanson & Brooks, 2010). The study discussed in this chapter had 82 participants, which is enough to detect whether and how nurse clarity of self, role, and/or system relate to nurse job satisfaction.

The variables of "nurse job satisfaction" and "clarity" are referred to scientifically as latent variables because they are invisible. However, when we quantify the self-reports of a specific population's sense of "clarity" of self, role, and system, as well as how satisfied they are in their jobs, these latent variables become visible. Further measurement then provides data about how clarity and job satisfaction relate to one another.

It was the hunch of the authors of this study that clarity of self, role, and system would predict nurse job satisfaction. If this relationship was found to be true in Jamaica, we would go on to study the relationship between clarity and job satisfaction in other countries. If the relationship between clarity and job satisfaction was found across countries, then deeper study could occur regarding context-specific variations of this relationship. This would then support training and operational changes to help improve job satisfaction in a way that aligned with the context of each country.

Latent variables have been found to vary from country to country. Some examples of latent variables that differ by country include entrepreneurial attitude (Stam, 2013), depression

(Rai, Zitko, Jones, Lynch, & Aray, 2013), bribery occurrence (Martin, Cullen, & Parboteeah, 2007), and life satisfaction (Vermuri & Costanza, 2006). Because contexts can differ significantly, each country's context must be considered individually if our aim is to improve clarity and nurse job satisfaction.

The Hunch: Where Measurement Begins

The initial hunch that nurses' clarity of self, role, and/or system predicts job satisfaction came about during research conducted by Persky, Nelson, Watson, and Bent (2008) which curiously reported that patients who reported they had the most caring nurses were actually cared for by the nurses with the lowest levels of job satisfaction. This counterintuitive finding was later contradicted by a replication study which found that the nurses whose patients reported they had the most caring of nurses were the nurses with higher levels of job satisfaction (Berry et al., 2013).

This was not a complete surprise, as insight into the earlier curious finding was provided by a 2011 study by Persky, Felgen, and Nelson. Like the 2008 study, this 2011 study was conducted in an organization practicing Relationship-Based Care® (RBC) (Creative Health Care Management, 2017; Koloroutis, 2004), and the study revealed that over the time that RBC was implemented, as nurses stepped more fully into their professional roles, they gained clarity of self, role, and system. Thirty-five units implemented RBC in five different groups of units, at different times, over 5 years. The implementation was done in "waves," and each wave was measured as a separate study to monitor changes over time.

Over the 5-year period, Primary Nursing (Manthey, 1980; Wessel & Manthey, 2015) was implemented, and this is where we began to develop our hunch that as nurses become clear in their roles, they evolved professionally. As a primary nurse gained clarity about the role of the nurse, using clinical and relational competence to impact patient care outcomes, nurse job satisfaction improved.

As the primary nurse gained clarity about the role of the nurse, outcomes improved.

In Primary Nursing in acute care settings, a group of nurses develop a process to assign patients whereby one professional nurse takes responsibility for a group of patients from admission through discharge on the unit. Continuity of nurse–patient assignments is a primary focus, but not a requirement, as the primary nurse works with associate nurses who are in essence co-assigned to the patient to support the primary nurse's plan of care for the patient. It is a case method style of care delivery.

According to Persky, Felgen, and Nelson (2011), in all groups studied while RBC was so new in the organization that its effects were not yet being experienced, it was found that satisfaction with scheduling was the foremost predictor of the nurses' satisfaction with Primary Nursing. However, upon the second or third year of RBC implementation, scheduling no longer predicted nurses' satisfaction with Primary Nursing at all. Once nurses began to experience the increased autonomy that comes with Primary Nursing, and once they saw the impact of their care and appreciated their ability to connect more deeply with patients and families, "professional knowledge of the nurse" replaced scheduling as the foremost predictor of nurses' satisfaction with Primary Nursing.

Once nurses began experiencing autonomy and a deeper connection to patients, professional knowledge replaced scheduling as their foremost satisfier.

Continuity of nurse–patient assignments would remain a goal of the organization, but this study revealed that continuity of assignment was not a requirement for nurses to connect to their patients. The findings from this Primary Nursing study led to this hunch: There is a relationship between clarity of self, role, and system and nurse job satisfaction. Persky, Felgen, and Nelson proposed that the nurses who had low job satisfaction in the early days of RBC implementations were frustrated with not yet having enough control over the social and technical aspects of their environments to make RBC operational (2011). However, as their clarity of role increased, they were able to figure out how to make RBC work, whether they felt supported by the system they were working in or not. Once these nurses began to experience clarity of role, they were more willing (and better able) to state what was and was not working in the system to care for the patient. This led to refinements of structures and processes to improve patient care. It also explained why nurses reported low levels of job satisfaction at the beginning of RBC implementation. They were frustrated because their increased clarity about the job led to the subsequent realization that there was a need to improve the systems in which they worked, both socially and technically, if they were to conduct patient care as needed.

They were frustrated because their increased clarity about the job led to the realization that there was a need to improve the systems in which they worked.

The Model

A structural model to examine whether clarity predicts job satisfaction was proposed for the study, based on the clarity work of Jayne Felgen and John W. Nelson (2016) and Nelson's work on job satisfaction (2001, 2013). The structural model in Figure 15.1 reveals what we desired to test: mainly, whether clarity of self, clarity of role, and clarity of system predict job satisfaction for nurses.

To ensure that we would get a true picture of whether clarity predicted job satisfaction, demographics related to nurse job satisfaction found in the literature (Nelson, 2013) were studied to ensure that demographics had no confounding effect on job satisfaction. The personal demographics examined in relationship to job satisfaction included each nurse's age, number of dependents, partner/marital status, household income, and whether the respondent was the primary source of family income. The professional demographics examined included each nurse's service line, years of experience on the same unit, level of education, number of continuing education hours, and number of hours worked per week. If relationships between any of these demographics and job satisfaction were found, they would be included, along with clarity, in the study of the predictors of nurse job satisfaction in Jamaica.

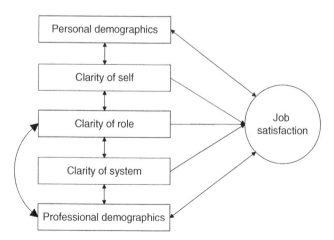

Figure 15.1 Proposed model to study whether clarity predicts nurse job satisfaction.

Using Theory to Explain the Model

Sociotechnical systems theory asserts that management and staff must work together to establish a good environment that is satisfying both socially and technically (Trist & Bamforth, 1951; Trist & Emory, 2005). Using this theory, Nelson (2001, 2013) defined nurse job satisfaction as "the feeling derived from perceiving that the social and technical aspects of the work environment are sufficient to perform the job" (p. 2). Since this theory proposes management work with staff to create satisfying work environments, Nelson proposes that nurses who have clarity of self, role, and system will be most efficient in collaborating with management for social and technical refinement of the work environment. According to Nelson, clarity of self is clarity at the individual level, clarity of role is at the group level, and clarity of system is at the organizational level (2013). A higher degree of clarity would make individuals more successful in getting their technical and social needs met in the system in order to execute the role expected of them. Fulfilling the social and technical dimensions of the work to a level that allows them to do their jobs well is what results in nurse job satisfaction (Nelson, 2013).

Role clarity also helps the interprofessional team leverage each other's expertise, which results in better patient outcomes (Ganann, Weeres, Lam, Chung, & Valaitis, 2019). One of the challenges to interprofessional practice is the lack of role clarity for nurses, especially when nurses understand their roles to be more like an assistant role than a professional role with unique and independent contributions to patient care (McInnes, Peters, Bonney, & Halcomb, 2017). Role clarity was found to be one of the most significant variables to explain the relationship between psychosocial stressors and productivity; in short, clinical and allied health professionals (e.g. doctors, nurses, pharmacists, dieticians, etc.) who had role clarity were able to navigate interdisciplinary relationships better to enhance team productivity (Ibrahim et al., 2019).

Measuring the Model

A measurement instrument to study clarity of self, role, and system had already been developed and psychometrically tested

(Nelson & Felgen, 2015). Job satisfaction was measured using the Healthcare Environment Survey (HES) which has been used to measure nurse job satisfaction throughout the world. The historical development of the HES is found in Nelson (2001, 2013), and extensive psychometric testing of the HES has been described by several authors (Nelson, Persky, et al., 2015; Nelson & Cavanagh, 2017). The 11 facets of the HES are shown here:

The four relational dimensions of nursing are:

1) Patient care delivery
2) Relationship with physicians
3) Relationship with co-workers (nurses and other co-workers)
4) Relationship with the unit manager

The seven technical dimensions of nursing are:

5) Autonomy
6) Workload
7) Professional growth
8) Executive leadership
9) Organizational rewards (including pay)
10) Staffing
11) Resources

Understanding the Context of Jamaica

It was deemed important to study nurse job satisfaction since no existing study of nurse job satisfaction in Jamaica could be found at the time of our study. It seemed especially important to study nurse job satisfaction when considering the large number of nurses who were leaving Jamaica for countries that pay higher wages (Kurowski et al., 2009). CARICOM (Caribbean Community), a large policy-making institution comprising 15 mostly English-speaking countries in the Caribbean, reports the concern that nurses leaving Jamaica for other countries negatively impacts the proficiency and safe practice of nursing staff through the loss of nurses who have years of experience working as nurses in the Caribbean (Kurowski et al., 2009). Migration has been reportedly due to salaries, career prospects, and education opportunities which all contribute to job satisfaction.

To study job satisfaction in Jamaica, it was proposed to the administration of a university hospital in the Southeastern portion of Jamaica to measure nurse job satisfaction and the relationship it had to clarity of self, role, and system. Administration reviewed the items in the proposed surveys to ensure they were appropriate for the context of Jamaica. Administration of the facility confirmed the items to be generally valid for the context of Jamaica, but some changes to wording would be required. For example, a manager of a clinical nursing care unit is not referred to as a "nurse manager," as it is in the United States, but rather as a "ward sister." (In recent years, due to an increase in men entering the field of nursing, the terms unit manager and ward manager have also become common.)

It should be noted that prior to use of the Healthcare Environment Survey (HES) to measure nurse job satisfaction in Jamaica, it was found that nurses in general, and broadly, had a consistently negative reaction to the term "Primary Nursing," and we found this same response by some managers in Jamaica, so we stopped using the term almost immediately. The negative reaction seemed to be due to nurses who had experience with a model of care that was given the name Primary Nursing but often had nothing to do with the principles or practices of Primary Nursing as defined by Manthey (1980), studied extensively by Nelson (2001), and as practiced in RBC organizations (and elsewhere) across the world. To avoid negative responses that would detract from the measurement of the *concepts* of Primary Nursing, we instead referred more generically to "professional patient care," which we measured with a tool we called the Professional Patient Care Index (PPCI). This tool measured what we know in the United States to be Primary Nursing as defined by Manthey (1980). The PPCI is one of the subscales or dimensions of the HES, an instrument developed by Nelson (2001, 2013).

Methods to Study Job Satisfaction and Clarity in Jamaica

This study was conducted in three stages:

Stage 1: First we examined the structure of nurse job satisfaction in Jamaica, specifically to see if there were both social and technical dimensions, or if instead, all factors of nurse job satisfaction were unidimensional. Finding two dimensions would support use of sociotechnical systems theory which posits that job satisfaction comprises both social and technical dimensions of work. Testing for two dimensions would be carried out using confirmatory factor analysis and fit indices which indicate whether the two-dimensional model is a good fit.[1]

Stage 2: We then studied the relationship between job satisfaction as measured by the HES and clarity as measured by the tool to measure clarity developed by Felgen and Nelson (2016), in the context of all nurse demographics. We did this in preparation for the final hierarchical regression analysis which would include only the variables that were found to be statistically significant.

Stage 3: Finally, we conducted a hierarchical regression using the variables found to have statistically significant relationships with job satisfaction.

Managing Disappointment with the Low Response Rate

To conduct the survey itself, paper surveys were handed personally by the researcher to all 364 nurses in the organization. The researcher spent 80 hours, with a research assistant and co-investigator, distributing and collecting the paper surveys. It quickly became evident that many participants were distrustful of an outside researcher from the United States collecting data, and at one point the researcher was even barred by the manager of the intensive care unit from entering the unit. Despite the explanation of the purpose of the study, consent forms being submitted with every survey, and a sealed envelope

---

1 A good fit for the root mean square error of approximation (RMSEA) is .06–.08 (Hooper, Coughlan, & Coughlan 2008). Other fit indices used included the comparative fit index (CFI), which should be above .90, with an index of .95 or greater being more desirable (Wang & Wang, 2012). The standardized root mean square residual SRMR should be less than .08 (Hu & Bentler, 1999).

with no name on the survey itself, only 82 of the 364 nurses responded to the survey—a disappointing 17% response rate. We had desired to have closer to 300 participants, which would have allowed us to test each measure using confirmatory factor analysis, but this number was considered adequate considering the extensive prior testing of the HES, its sound theoretical underpinnings, and a sample size review from psychometric testing experts Tabachnick and Fidell (2007).

While it was disappointing to receive such a low response rate, the 82 respondents ultimately provided enough data to test the relationship between clarity and job satisfaction. It was also helpful to have tested our nurse job satisfaction tool, the Healthcare Environment Survey, in numerous other hospitals in numerous other countries prior to this study (Nelson, Persky, et al., 2015). Because the tool had been used in so many countries, the HES could be tested using a method called parceling of data[2] which is allowed if the measurement instrument has already shown consistency in behavior across many studies.

Results on the Social and Technical Dimensions of Nurse Job Satisfaction in Jamaica

Confirmatory factor analysis revealed that all survey items loaded into the appropriate 11 factors of the HES. Some items had factor loadings lower than the desired .3, but considering the small sample size and past performance in other studies of the HES to measure nurse job satisfaction, it was decided that these results were adequate to proceed with evaluation of the two-factor solution using parceling of data. The 11-factor solution from the confirmatory factor analyses is noted in Figures 15.2 and 15.3.

Note that only 9 of the 11 factors are included in the model in Figure 15.2 and all 11 factors are noted in Figure 15.3. This is because the "staffing/scheduling" and "satisfaction with resources" subscales were new and needed to be studied

---

2 Parceling data is used when there are a large number of factors. It involves bundling items from a subscale into a mean score. This enables a more parsimonious model with more stable parameter estimates than factor models that include individual items (Nelson, 2013; Wang & Wang, 2012).

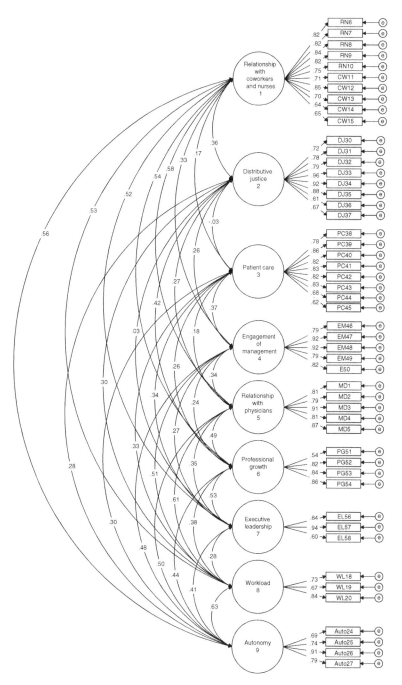

Figure 15.2 Confirmatory factor analysis for 9 of the 11 items from the Healthcare Environment Survey (HES).

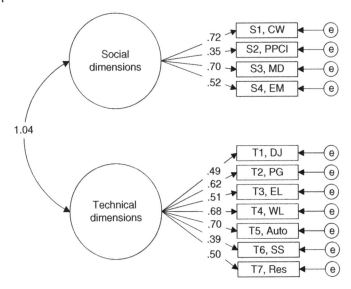

Figure 15.3 Two-dimensional model of nurse job satisfaction with 11 factors and factor loadings.

separately before combining with the 9 other subscales. For information on the factor loadings for satisfaction with staffing/scheduling and resources, see Appendix T.

When looking to see whether the 11 factors of nurse job satisfaction loaded into a two-dimensional model of nurse job satisfaction, it was confirmed that there were 4 social factors and 7 technical factors that contributed to nurse job satisfaction in Jamaica. All factor loadings were above .3, which was the minimum set for a factor to "belong" to the model.

The social dimensions, in order of importance, were:

1) Relationship with coworkers (factor loading .72),
2) Relationship with medical doctors (factor loading .70),
3) Engagement with the unit manager (factor loading .52), and
4) Professional patient care (Primary Nursing) (factor loading .35).

The technical dimensions, in order of importance, were satisfaction with:

1) Autonomy (factor loading .70),
2) Workload (factor loading .68),
3) Professional growth (factor loading .62),

4) Executive leadership (factor loading .51),
5) Resources (factor loading .50),
6) Distributive justice/professional rewards (factor loading .49), and
7) Staffing/scheduling (factor loading .39).

This 11-factor, two-dimensional model had good fit indices, including an RMSEA of .07, a CFI of .95, and an SRMR of .06. The two-dimensional model with 11 factors and factor loadings are noted in Figure 15.3.

Definitions for the model depicted in Figure 15.3 are as follows:

- S1, CW = Satisfaction with relationship with coworkers;
- S2, PPCI = Satisfaction with professional patient care
- S3, MD = Satisfaction with relationship with medical doctors
- S4, EM = Satisfaction with engagement with unit manager
- T1, DJ = Satisfaction with distributive justice (professional rewards such as pay)
- T2, PG = Satisfaction with professional growth
- T3, EL = Satisfaction with executive leadership
- T4, WL = Satisfaction with workload
- T5, Auto = Satisfaction with autonomy
- T6, SS = Satisfaction with staffing/scheduling
- T7, Res = Satisfaction with resources

Results on the Relationship of Role Clarity and Demographics to Nurse Job Satisfaction in Jamaica

In Jamaica, clarity of self and clarity of system did not have a statistically significant relationship with job satisfaction, though clarity of role did, and the only demographic that was found to predict nurse job satisfaction was what service line the nurse worked on. When these two variables—clarity of role and service line—were then entered into a hierarchical regression equation, it was found that service line predicted 22.6% of the variance of nurse job satisfaction, and that clarity of role predicted 6.9%. When combined, they explained 29.5% of what predicts nurse job satisfaction, both socially and technically, for the nurses in Jamaica who responded to the survey.

Application of the Findings

The findings of this study were presented to all nurse managers in the organization to help them understand the state of nurse job satisfaction socially and technically. The factor analysis in this study, as noted in Figure 15.3, revealed that nurses' relationships with coworkers and medical doctors were the most important social variables contributing to satisfaction, and that autonomy and workload were the most important technical variables contributing to satisfaction. Using a 7-point Likert scale, with higher numbers indicating greater satisfaction, it was noted that for the nurses in their organization, mean scores on three of four of these important variables were above the midpoint of 4.0, indicating general nurse job satisfaction. Only satisfaction with the relationship with doctors was below the midpoint of 4.0, indicating overall dissatisfaction with the nurse–physician relationship. This finding must be a priority finding considering it ranks highest in the factor analysis (meaning it is very important to nurse job satisfaction) but has the lowest mean score in satisfaction. Thus, from an analytic point of view, the nurse–physician relationship must be a target variable for process improvement within operations.

The nurse–physician relationship must be a target variable for process improvement within operations.

It was also interesting to find in this study that clarity of role was the sole predictor of job satisfaction among the three aspects of clarity studied. Historically, organizations have reported clarity of system as more predictive of job satisfaction (Nelson, Nichols, & Wahl, 2017). For example, you may recall that in the study from Chapter 10, clarity of system predicted 42% of job satisfaction.

It is likely that the emergence of role clarity as a predictor of job satisfaction in this current study is directly due to the unsatisfactory relationships identified between nurses and physicians. It is common for us to discover that nurses who are reporting low satisfaction in role clarity are dissatisfied because they perceive that physicians do not understand the different levels of the role of the nurse, or the independent, interdependent, and dependent dimensions of the nurse's role. Sometimes the physician will ask the nurse to do tasks that are either too technical or too menial because they do not know the full scope of what the professional nurses they work with are licensed to do or they do not understand the differences between the role of the professional nurse and the role of a nursing assistant. Nurses and physicians may also experience conflict over something they are both licensed to do. Conversely, nurses who score themselves higher in role clarity generally have greater confidence in communicating to physicians what is and is not within the nurses' scope of practice. The nurse with clarity can also communicate confidently with people in other roles, such as respiratory therapists or occupational therapists, with whom the nurses' training sometimes overlaps. Role clarity helps to clarify not only what a person's professional licensure or certification allows, but how the organization's policy determines who is to carry out what tasks within the institution.

To understand the relationship with coworkers (other than physicians) in more detail, the five items dealing with nurses' relationships with other nurses were examined separately from nurses' relationships with coworkers more generally, and it was found that satisfaction with nurses' relationship with nurses and nurses' relationship with coworkers more generally was about the same, with mean scores of 4.37 and 4.39, respectively.

It was especially interesting to note the wide range of scores given by respondents, with minimum scores usually being approximately a score of one, and the maximum being a score of seven. This is fairly typical in the studies we have done on nurse job satisfaction across the world. This variance of experience among staff members was further supported by the standard deviation greater than 1.0 for all subscales.

It is important to understand which nurse job satisfaction factors were above and below the midpoint of 4.0, indicating satisfaction or dissatisfaction, respectively. In Table 15.1, the first eight factors scored a mean of four or above, meaning staff members were satisfied with them.

As you can see, staff members were not satisfied with the last five factors in Table 15.1. When considering all job satisfaction factors in the Healthcare Environment Survey (HES) as a group, the nurses' mean score was 4.09, meaning they were slightly more satisfied than dissatisfied with their jobs overall. It is noted that the standard deviation of all the factors is above 1.0, indicating that nurses in this study were having a wide range of experience as it relates to nurse job satisfaction.

Nurses in this study were having a wide range of experience as it relates to nurse job satisfaction.

Table 15.1 Job satisfaction factors for nurses in Jamaica, from most to least satisfactory.

Factors and total scores	N	Min	Max	Mean	SD
Professional patient care	82	1.50	7.00	5.25	1.21
Clarity of role	82	1.17	6.00	4.77	1.04
Engagement with management	82	1.00	7.00	4.57	1.64
Autonomy	82	1.00	7.00	4.49	1.54
Relationship with coworkers	82	1.20	7.00	4.39	1.28
Relationship with nurses	82	1.00	6.60	4.37	1.35
Workload	82	1.00	6.67	4.32	1.48
Executive leadership	82	1.00	6.67	4.21	1.39
Staffing/Scheduling	82	1.00	7.00	3.99	1.66
Professional growth	82	1.00	6.50	3.84	1.51
Relationship with physicians	82	1.00	6.40	3.73	1.39
Resources	82	1.00	6.67	3.41	1.55
Distributive justice	82	1.00	6.00	2.61	1.32

Definitions: *N* number of nurses who are in sample, *Min* minimum score on the 7-point Likert scale, *Max* maximum score in the 7-point Likert scale, *SD* standard deviation.

In many Caribbean countries, resources, opportunities for career advancement, and professional growth have all been described as lacking among nurses (Salmon, Yan, Hewitt, & Guisinger, 2007), and these crucial lacks undermine job satisfaction. The findings also highlight the low mean satisfaction scores of factors relating to the technical dimensions of nursing.

Managers attended a presentation of these results and reported these findings helpful. It is common to have managers say, after hearing results, "Nothing here surprises me, but it is helpful to see it in numbers and graphs." This provides the managers with an objective tool to use in conversations with both staff and administrators. Having data that is trusted and that resonates with the lived experience of the staff allows several things: (a) it validates what is suspected, (b) it provides a report that moves subjective assertions about what is perceived to an objective report of "what is," and (c) it details a map for change with priorities for action (e.g. nurse–physician relationships). The results also helped clarify some of the scheduling and management issues that were happening in some units, which helped to some degree with staffing.

The authors of this report discussed why professional patient care/Primary Nursing had the lowest factor loading, meaning it had the weakest connection to the construct of nurse job satisfaction among all 11 social and technical factors measured. This is in contrast to all of the other countries where the HES has been used to measure nurse job satisfaction, which have shown that Primary Nursing, as measured by the PPCI, consistently had among the top three highest factor loadings (Nelson & Cavanagh, 2017; Nelson, Persky, et al., 2015). Since we had removed the troubling moniker "Primary Nursing," we knew it had to be something else. The author from Jamaica was not at all surprised by this finding, and she explained it simply: "It takes a village to provide care." Items in the PPCI were measuring, in part, the extent to which one individual was taking primary authority for the care of a set of individual patients—something that ran directly counter to the spirit of interdependency of the whole "village." In Jamaica, patient care is a team effort, as is consistent with the overall culture in Jamaica, where it is expected that all people in a community will care for one another. From the Jamaican point of view, if teamwork is not present, if managerial support is not present, if resources are not present, then patient care does not happen.

It may be tempting for those who live in cultures where autonomy and individuality are prized to assert that this team in Jamaica should be encouraged to move from an approach of serving the patient as a village to a more individualistic format of care delivery where, for example, there is a primary physician and a primary nurse. However, this research and subsequent interpretation suggests that an approach to care delivery design that amplifies the deeply engrained and perhaps cherished characteristics of the culture in which the care is taking place could and should be examined more closely. This is not only to learn about the specifics of various cultures but to look for global lessons about caring for others as a village. This illustrates why theories and measurement instruments must both be tested each time they are used within a new context, and the findings must be interpreted by those who live and work in the context being measured.

16

Testing an International Model of Nurse Job Satisfaction to Support the Quadruple Aim

John W. Nelson, Patricia Thomas, Dawna Cato, Sebahat Gözüm, Kenneth Oja, Sally Dampier, Dawna Maria Perry, Karen Poole, Alba Barros, Lidia Guandalini, Ayla Kaya, Michal Itzhaki, Irit Gantz, Theresa Williamson, and Dominika Vrbnjak

The Triple Aim was initially proposed by the Institute for Healthcare Improvement (IHI) to improve the patient experience and population health and decrease healthcare costs (Berwick, Nolan, & Whittington, 2008). The Quadruple Aim proposes that improvement of the employee's experience of work, which in the study described in this chapter is the experience of the nurse, is an essential prerequisite to achieving the Triple Aim (Bodenheimer & Sinsky, 2014). We know from the Jamaican study described in Chapter 15 that clarity of role impacted nurse job satisfaction, and we know from other studies that clarity of system has also been shown to predict nurse job satisfaction (Nelson, Nichols, & Wahl, 2017). We also know from numerous studies that nurse job satisfaction can be improved by increasing nurses' clarity of role (Nelson & Felgen, 2015; Nelson et al., 2017), clarity about the system in which they work (Nelson et al., 2017), caring for self (Nelson & Felgen, 2015), and the caring of the unit manager (Bolima, 2015). With the exception of our study in Jamaica, the empirical evidence regarding the importance of these four work experience factors has come solely from research done in the United States. It is the hypothesis of the authors, however, that these work experience factors would resonate with nurses globally. The goal of the study reviewed in this chapter was (a) to examine the relationships that the work experience factors of clarity of role, clarity of system, caring for self, and the care of the unit manager have with nurse job satisfaction, across eight countries, and (b) to examine the extent to which these four work experience factors *predict* nurse job satisfaction and the associated outcomes of decreased turnover and decreased sick time of nursing staff.

Prior to this study, most of the work done to address the Quadruple Aim has focused on the single variable of clinician burnout (Anandarajah, Quill, & Privitera, 2018; Havens, Gittell, & Vasey, 2018; Privitera, 2018). There were other single variables that were studied in relation to the Quadruple Aim, including callousness (Anandarajah et al., 2018), resilience (Annandarajah et al., 2018), well-being (Brown-Johnson et al., 2019; Jacobs, McGovern, Heinmiller, & Drenkard, 2018), engagement (Havens et al., 2018; Jacobs et al., 2018), and job satisfaction (Longbrake, 2017). The authors of this chapter propose that a multivariate model of the experience of work, as opposed to a study examining a single variable, will provide for a more nuanced and actionable explanation of what might be done to improve the work experience, in pursuit of advancing the Quadruple Aim. Bodenheimer and Sinsky (2014) advocated for the Quadruple Aim and used the term "workforce" to describe healthcare staff. Figure 16.1 illustrates a model that could be used to illustrate the Quadruple Aim.

As the model indicates, the three variables hypothesized to predict the Triple Aim of an improved healthcare system are (a) improved population health, (b) reduced costs for healthcare, and (c) an improved patient experience. Bodenheimer and Sinsky (2014) realized there was one more variable that could predict all three of those variables: the experience of work of the healthcare workforce. The theory behind the Quadruple Aim is that the more gratifying the experience of work for healthcare workers is, the more likely it is that population health is improved, costs for healthcare are reduced, and the patient experience is improved. In short, the experience of work for the healthcare workforce becomes the engine that drives the other three variables.

To create the study outlined in this chapter, the authors used findings from previous work on the effects of clarity and caring in nursing and prior work from the mining industry on the importance of workers experiencing

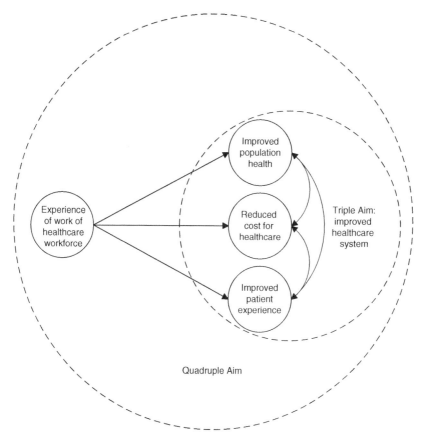

Figure 16.1 Model for the Quadruple Aim.

satisfaction in both the social and technical aspects of their work. Examination of these theoretical constructs led us to develop a theoretical framework for the experience of work: It was theorized that nurses' clarity of role and system, their caring for self, and the care they feel from their managers would all contribute to their job satisfaction. These four things would create a positive experience of work for the employee, thereby fueling the engine behind the Triple Aim, and bringing organizations and individuals closer to satisfying the Quadruple Aim.

The Four Goals of Our Study

Goal 1: Create a model of the experience of work measuring the effects of clarity of role, clarity of system, caring for self, and caring of the nurse manager that would be valid and reliable across eight countries.

Goal 2: Validate clarity and caring as constructs that predict nurse job satisfaction across eight countries.

Goal 3: Examine the relationship between nurse job satisfaction and the associated outcomes of nurse turnover and nurse sick time across eight countries.

Goal 4: Use structural equation modeling (SEM) to test a model of the experience of work for the healthcare workforce for validity using several model fit indices across eight countries.

Methods

This was a descriptive, nonexperimental study that used factor analysis to study the factor structures of the instruments used and SEM to study the effects on job satisfaction of nurses experiencing clarity of role, clarity of system, caring for self, and the caring of their nurse managers.

Sample and Setting

We used a sample of 11 mid-sized (250–500 beds) or large (>500 beds) hospitals in 8 countries: Brazil, Canada, China, Israel, Scotland, Slovenia, Turkey, and the United States. Eligible respondents from each hospital were limited to nurses involved in direct patient care and considered employed by the hospital. Exclusion criteria included travel nurses, advanced practice and managerial nurses in administrative positions without direct patient care, temporary staff, those who have not completed orientation if recently hired, and those on leave (e.g. maternity or sick leave). Several hundred nursing assistants (NAs) who met the inclusion criteria were included, as NAs are integral to care delivery in Brazil, Scotland, Slovenia, and the United States. Support for the decision to include NAs was also found in research showing that the factors for job satisfaction are the same for nurses and NAs (Nelson & Cavanagh, 2017). There were a total of 7,617 nursing staff who met this criteria and were invited to participate.

Nurse leadership in each organization submitted the study protocol to their organizational institutional review board (IRB). Each organization submitted a protocol that was tailored to their organization's standard method of communicating with staff to invite all eligible nurses to participate. To minimize the perception of, and potential for, coercion, all unit/ ward leaders were instructed in the confidentiality of the study and a consent form accompanied each survey to assure potential respondents that their response was voluntary. After IRB approval, each organization initiated the study.

Theoretical Framework

Each of the measurement instruments used in the study had its own theory underpinning it. The theoretical framework for caring on which the final model is based is the 10 Caritas processes of Watson's Theory of Transpersonal Caring (2008a), which asserts that if 10 specific processes of caring are enacted, caring will be perceived by the person the process was enacted upon, wheather it was enacted upon one's self or another person. Once again, the 10 Caritas processes of caring, as proposed by Watson, are:

1) Cultivating the practice of loving kindness and equanimity toward self and others. Loving kindness includes listening to, respecting, and identifying vulnerabilities in self and others.
2) Being authentically present: enabling, sustaining, and honoring faith and hope which is future-oriented and includes self-discovery.
3) Cultivating one's own spiritual practices and transpersonal self, going beyond ego-self.

4) Developing and sustaining a helping-trusting caring relationship.
5) Being present to, and supportive of, the expression of positive and negative feelings.
6) Creative use of self and all ways of knowing as part of the caring process; engaging in the artistry of Caritas (caring). At the core here is creative problem solving.
7) Engaging in genuine teaching-learning experience that attends to unity of being and subjective meaning: attempting to stay within others' frame.
8) Creating a healing environment at all levels.
9) Administering sacred acts of caring-healing by tending to basic needs.
10) Opening and attending to spiritual/mysterious and existential unknowns of life-death. This is belief in the impossible (miracles), even when others may assert doubt.[1]

In this study, Scotland chose their own theory of caring, derived from the National Health Service (NHS) of Scotland (The Scottish Government, 2010). They used a different theory after trialing the 10-item Watson tool and finding it was not a valid measure of caring for the context of Scotland. They also attempted to develop a measure of caring using Swanson's Theory of Caring (2008), but that was also found to not be valid in the context of Scotland. The NHS theory of caring was derived from the 2010 Healthcare Quality Strategy for NHS, which was based on the 7 Cs: Caring, Compassion, Communication, Collaboration, Clean Environment, Continuity of Care, and Clinical Excellence. Development of the instruments used to measure caring in Scotland is described in detail in Chapter 17 of this book.

The theoretical underpinning for the "caring for self" element of the study comes from Nelson et al. (2017), who showed that caring for self predicts nurse job satisfaction. The theoretical underpinning for the benefits of having a caring unit manager comes from Bolima (2015), who showed that having a caring manager predicts nurse job satisaction. The theoretical framework for clarity used in the study was from Felgen's Theory of Clarity (Felgen & Nelson, 2016). According to Felgen, there are three dimensions of clarity as it relates to the

---

1 Nelson, DiNapoli, Turkel, & Watson, 2011.

professional role of the nurse: clarity of self, clarity of role, and clarity of system (Felgen & Nelson, 2016). Clarity is foundational for enacting the professional role of the nurse (Felgen & Nelson, 2016). It has been shown that nurses who have clarity, as defined by Felgen, have higher levels of job satisfaction (Nelson & Felgen, 2015; Nelson et al., 2017).[2]

In addition to these theories on caring and clarity, we considered the importance of sociotechnical systems (STS) theory, which proposes that it is critical for management to examine and refine both the social and technical dimensions of workers' jobs for better employee and organizational outcomes (Trist & Bamforth, 1951; Trist & Emery, 2005). While STS theory comes from the field of mining, it has been shown to be applicable to healthcare (Nelson, 2001, 2013), and all of the researchers in this study agree that it is valid in this study. As in mining, good outcomes for both nursing professionals and their patients require nurses to have the proper social and technical skills to perform their jobs (Nelson 2013; Nelson et al., 2017; Nichols, Hozak, & Nelson, 2016). Clinicians also need to be informed about how systems of care work and how being *un*informed of how systems of care work has historically been a primary reason for failing to achieve the Triple Aim (Berwick et al., 2008).

Measurement Instruments and a Model of Measurement

The compilation of instruments used in this study included the 57-item Healthcare Environment Survey (HES) with the addition of one global question on job satisfaction to use for criterion validation and one qualitative question to evaluate what was most important for job satisfaction. There were also 15 items on clarity of role and system, 10 items on caring for self, 10 items on caring of the unit manager, and 14 demographic questions. Scotland's measures for caring for self and caring of the unit manager included only seven items each, but other than that difference, they used the same instruments. In

_____

2 Felgen's Theory of Clarity is an outgrowth of her work with Mary Koloroutis on the "5 Cs"—the five conditions for engaging in change—which were introduced in 2007 (Felgen & Koloroutis, 2007).

total, there were a maximum of 108 items presented to potential respondents. All questions, except the demographic items and the qualitative item, used a 7-point Likert scale from strongly disagree (1) to strongly agree (7).

The HES was selected from among 96 other nurse job satisfaction measurement instruments (Nelson, Gallagher, Cummings, Kaya, Nichols, & Thomas, 2018) as it was the only instrument demonstrating that measuring the nurses' relationship with the patient explained a large variance of nurse job satisfaction (Nelson, Persky, et al., 2015). It had also been found to be psychometrically sound in Jamaica, Scotland, Turkey, and the United States (Gözüm, Nelson, Yildirim, & Kavla, 2021).

As mentioned in previous chapters, the HES measures 11 factors of nurse job satisfaction and was developed using concepts of STS theory, which speaks to the importance of a positive experience in both the social and technical dimensions of work. Social dimensions within the HES include satisfaction with relationships with other nurses/coworkers, doctors, and managers, and satisfaction with caring for the patient. Technical dimensions include satisfaction with workload, autonomy, professional growth, scheduling, resources, executive leadership, and distributive justice, which is defined as satisfaction with rewards when considering one's level of effort, education, and experience.

The Caring Factor Survey—Caring for Self (CFS-CS) and the Caring Factor Survey—Caring of Manager (CFS-CM) both comprise 10 items using a 7-point Likert scale and were selected to measure caring in every context except Scotland. They are derivative measures of the Caring Factor Survey (CFS), which is based on Watson's Theory of Transpersonal Caring, which was co-developed to measure all 10 processes of Caritas (caring) (DiNapoli, Turkel, Nelson, & Watson, 2010). These measures have been psychometrically tested in multiple countries and found to be valid (Lawrence & Kear, 2011; Nelson, Itzhaki, et al., 2011; Olender & Phifer, 2011).

The clarity measure by Felgen and Nelson (2016), based on Felgen's Theory of Clarity (2016), is a 15-item instrument that uses a 7-point Likert scale to measure clarity of role and system. Based on the theoretical framework proposed by Felgen, it was selected because it has been shown to be psychometrically sound (Felgen & Nelson, 2016). Four items we had previously used to measure clarity of self had proven to be less effective than we had hoped in previous analyses, so for this study and several others we eliminated those items.

Outcome items to be measured included turnover and sick time at the unit level. Data on these metrics was provided by the organizations. Model 1, designed to find out whether clarity and caring predict job satisfaction and the subsequent outcomes of nurse turnover and sick time, is depicted in Figure 16.2.

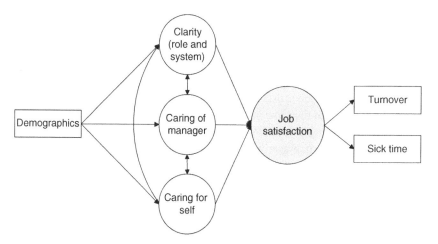

Figure 16.2 Model 1: a model to research the effects of nurse job satisfaction on turnover and sick time.

Order of Operations of the Study

Step 1: Data was to be collected using either paper or an electronic link; whichever was more operationally feasible for the organization. The survey was available to potential respondents for three weeks. Accommodations were made if organizations needed more time. A response rate of 40% was deemed adequate to represent the sample (Kramer et al., 2009). Each facility decided the best time to launch the study.

Step 2: Once the surveys closed, the data was downloaded from the Healthcare Environment Data and Survey Software by the primary investigator, Dr. John W. Nelson, who examined the dataset for missing data and potential bias (Nelson et al., in review). There were 2,579 surveys returned which represented a 33.9% response rate. After screening returned surveys for missing data, there were 2,046 surveys used for analysis which represented a 79.3% completion rate.

Step 3: Exploratory and confirmatory factor analyses of the data from the 57-item job satisfaction survey were done to see whether the primary measurement tool had good model fit and whether all items loaded into their respective subscales. The same procedure was then conducted for the caring and clarity tools.

Step 4: Structural equation modeling (SEM) was then used to study how the latent variables of job satisfaction, caring, and clarity related to one another. The relationship between clarity and job satisfacition was studied first, since all eight countries used the same instruments to measure them. In a second model using SEM, the variables of caring for self and the caring of the unit manager were added to clarity and job satisfaction, but for only six of the eight countries due to Scotland using a different theory of caring and Israel not measuuring caring at all due to limits on the number of items nurses were able to respond to. A separate analysis was conducted for Scotland, replicating the SEM analysis and will be published in a journal article after the publication of this book.

Step 5: The next step was presentation of results by the lead investigator to co-investigators and leaders from each partici-pating facility. It was essential for researchers and leaders to see the results for each of the variables from within their own facilties. The hope was that they could use their own organiza-tion-specific data to improve the social and technical aspects of the work environment to promote caring for self, caring of the unit managers, and clarity of role and system. Operational refinement was the incentive for the organizations to participate in the study in the first place, and each participating facility imple-mented action plans after reviewing their results. Results and action plans from each of the facilities will be written about at length in separate manuscripts for publication.

Step 6: Dissemination of findings in 2020 included writing articles for journals that provided much more detail than what has been described in this chapter. In addition, presentations were planned for dissemination of results at conferences regarding how the data was used operationally in the various countries to make changes to improve the work experience of nurses.

Emails and conference calls continued throughout 2020 as the international group of researchers planned for next steps to replicate and extend this research. An addditional three countries replicated this study with plans to analyze and report whether the results are the same as in our study.

Simplifying the Model

Due to the extensive information in the model and the desire to utilize the information operationally, results from the model were reported in phases, first establishing how clarity predicts job satisfaction and ending with an examination of the extent to which caring predicts both clarity and job satisfaction.

What We Needed to Understand About Job Satisfaction and Clarity Before We Added Caring

The number of survey items measuring job satisfaction and clarity was reduced after finding covariance (overlap or redundancy) among the items. To take care of this issue, the HES was reduced from a 57-item instrument with 11 dimensions to a 19-item instrument with 6 dimensions or subscales. This reduction was due to both redundancy and the fact that some items were found to be more precise in their measurement of nurse job satisfaction. The 15-item clarity instrument with two dimensions was reduced to a 4-item instrument with a single subscale.

We were glad to find that the Healthcare Environment Survey (HES) was a psychometrically sound measure in all 8 countries and that we could reduce it to 19 items without sacrificing validity. This not only makes this 19-item version of the HES the first global measure of nurse job satisfaction, but one that is concise and explains 82% of the variance of nurse job satisfaction. This was most interesting to find when considering that the 57-item HES explained only 78% of the variance of job satisfaction, which was already an explanatory measure, but lacked brevity. This not only gives us a good instrument to use going forward, it also reveals that nurses across the world think largely the same way about what is important for job satisfaction. This similarity

This not only gives us a good instrument to use going forward, it also reveals that nurses across the world think largely the same way about what is important for job satisfaction.

of thinking across the eight countries was further confirmed through invariance testing[3] and is described in detail elsewhere (Nelson et al., in review).

The first round of the study, which included just clarity and nurse job satisfaction, yielded a very helpful finding that we were able to take forward productively into subsequent rounds of the study. It also validated the study finding from Jamaica regarding clarity and job satisfaction described in Chapter 15. We now refer to the remaining six subscales of the HES as "the Big 6" international factors that explain nurse job satisfaction. Nurses across the world experience satisfaction when:

1) Their education and experience are rewarded (variable = "professional rewards")
2) Their managers listen to their ideas and concerns (variable = "communication with the unit manager/participative management")
3) They feel like they have a relationship with their patients (variable = "patient care")
4) Their co-workers are friendly, helpful, and respectful (variable = "relationship with co-workers")
5) They have opportunities to grow and learn professionally (variable = "professional growth")
6) They are able to provide care in the way they believe it should be carried out (variable = "autonomy")

The 5 subscales from the original 57-item HES that were cut out of the 19-item version were satisfaction with resources, scheduling, relationships with doctors, executive leadership, and workload. Each of these five subscales explained 1% or less of what comprises nurse job satisfaction and thus did not contribute enough to be included in the Big 6 (Nelson et al., in review).

We used exploratory and confirmatory factor analyses to reduce the 15-item clarity measure to 4-items. When studying how clarity predicted nurse job satisfaction, we found that clarity predicted all of the "Big 6" nurse satisfiers except "professional rewards." Clarity had the biggest impact on nurse satisfaction as

Clarity had the biggest impact on nurse satisfaction as it pertained to issues of patient care and autonomy.

_____

3 Measurement invariance testing is a statistical test to ensure that an instrument (e.g. a test, survey, or questionnaire) measures the same construct(s) in the same way across subgroups of respondents (Wang, Chen, Dai, & Richardson, 2018).

it pertained to issues of patient care and autonomy, which each had factor loadings of .72.

Using structural equation modeling, we found that our model of clarity and job satisfaction had excellent model fit in this eight-country study. See Figure 16.3 and Table 16.1.

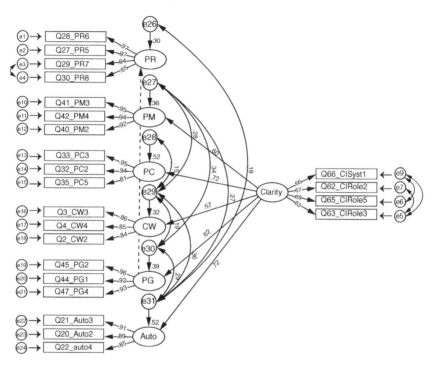

Figure 16.3 Model 1 for job satisfaction and clarity internationally.

Table 16.1 Indicators of how well the model fit what we intended to measure.

Fit indices	Desired fit indices, HES and clarity	Final fit result
(A)GFI	Greater than .90, with closer to 1.0 being preferred for superior fit	.947
CFI	.90 or greater but closer to .95 for superior fit	.983
SRMR	Less than .05 indicates fit	.0340
RMSEA (Lo90-Hi90)	Less than .06 indicates fit	.041 (.039–.044)
PCLOSE	Greater than .05 indicates fit	1.00

Respecifying the Model to Include Caring

The final model included only six of the eight countries because Israel's researchers used an abbreviated questionnaire with the aim of increasing the response rate of nurses at a time when many surveys were being distributed by the organization, and Scotland used different instruments to measure caring of the unit manager and caring for self.

Covariances for the instrument used to measure caring for self and the caring of the unit manager were identified. Due to the overlap or redundancy found in the instrument, both instruments were reduced from 10-items to 6-items. For more detail about the retained items, see Appendix U.

The respecified model is illustrated in Figure 16.4 and an assessment of its fit is found in Table 16.2.

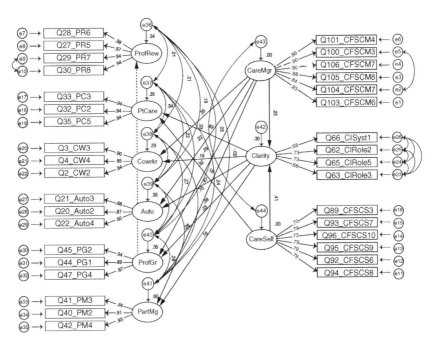

Figure 16.4 Model 2 to measure the relationship of job satisfaction, clarity, and caring internationally.

Table 16.2 Indicators of how well the respecified model fit what we intended to measure.

Fit indices	Desired fit indices, HES and clarity	Final fit result
(A)GFI	Greater than .90, with closer to 1.0 being preferred for superior fit	.928
CFI	.90 or greater but closer to .95 for superior fit	.977
SRMR	Less than .05 indicates fit	.035
RMSEA (Lo90-Hi90)	Less than .06 indicates fit	.038 (.036–.040)
PCLOSE	Greater than .05 indicates fit	1.00

Results from Model 2

Analysis revealed that caring for self predicted clarity, with a predictive value of .41, and the caring of the unit manager predicted clarity with a predictive value of .25.[4]

Further examination of the second model revealed that the only factor of job satisfaction predicted by caring for self was care for the patient, with a predictive value of .18, while the caring of the unit manager predicted four factors of nurse job satisfaction:

1) Satisfaction with communication with the manager was most strongly predicted by nurses' perception of caring of their unit manager, with a predictive value of .70.
2) Satisfaction with coworker relationships, which dealt primarily with teamwork, was a distant second, with a predictive value of .35.
3) Satisfaction with professional growth was third, with a predictive value of .32.
4) Satisfaction with autonomy was fourth, with a predictive value of .22.

---

4 The technical term for this number is the coefficient, and the symbol used to note it is R^2, which indicates a regression equation within the complex structural equation model (SEM). There is more information in Figure 16.4, and those who are trained in how to use SEM will be fascinated with how much this model informs about the Profile of Caring (see Chapter 6 for more on Profile of Caring).

The only one of the "Big 6" international factors that explain nurse job satisfaction that was *not* predicted by caring of the unit manager was satisfaction with professional rewards. This finding requires deeper inquiry regarding why caring of the manager did not predict satisfaction with rewards.

How Job Satisfaction Relates to Turnover and Sick Time

Seventy-two units in four of the eight countries were able to report outcomes data at the unit level. Due to the varied definitions and methods of data collection from each of the participating hospitals, however, we ended up with data on sick time for only 39 units and data for turnover on only 61 units. With this small sample size, we used a Pearson's correlation to examine the strength of the relationships. We found that job satisfaction had a negative relationship with both turnover ($r = -.13$, $p = .332$) and sick time ($r = -.255, p = .117$), indicating that as job satisfaction increased, turnover and sick time both decreased. We were not able to report statistically significant results, however, even if we adjusted to a less strict alpha and larger effect size in a post hoc power analysis. Still, we remain confident that the direction of the relationship is in the direction we proposed—specifically, that increased nurse job satisfaction correlates to decreased turnover and sick time. A larger study is required to provide a more reliable argument that this is indeed true.

Focus on Turnover

Our study demonstrates that a good experience of work can decrease turnover, and the literature supports our findings. Poor job satisfaction has been found to affect turnover (Griffeth, Hom, & Gaertner, 2000; Halter et al., 2017), as has participative management (Griffeth et al., 2000), promotional opportunities (Griffeth et al., 2000), teamwork (Griffeth et al., 2000), role clarity/ambiguity (Halter et al., 2017), and work climate (Halter et al., 2017). All of these constructs are included in our model of work experience and thus provide additional support to the findings of our study.

Turnover has been shown to have both direct and indirect costs (Li & Jones, 2013). Costs associated with turnover are tied to vacancy rates, termination, hiring, marketing and recruiting, orientation and training, and decreased productivity (Li & Jones, 2013). Turnover costs range internationally from $10,098 to $88,000 per registered nurse (RN) (Li & Jones, 2013). As we have shown that a positive experience of work (inclusive of caring for self, the caring of the unit manager, clarity, and job satisfaction) decreases turnover, we see also that it reduces costs for healthcare overall, as Figure 16.5, which you may recall from the beginning of this chapter, suggests.

Turnover rates vary from country to country. Understanding the global rates of turnover may help this current group of

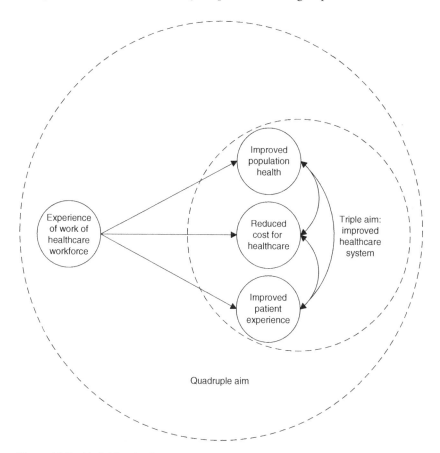

Figure 16.5 Model for the Quadruple Aim.

researchers, as an international cooperative, find funding to support countries with the highest turnover rates to improve their work environments.

Sick Time

The cost of sick time for a 5.6% sick rate in three regional health authorities in Canada that included almost 37,000 employees working just over 36,000,000 hours was $52,621,625 for 1 year (Gorman, Yu, & Alamgir,. 2010). Our study shows that a positive experience of work decreases sick time, which again reduces costs for healthcare. But absenteeism was also found to have a negative relationship with patient satisfaction (Duclay, Hardouin, Sebille, Anthoine, & Moret, 2015), which means that work experience/job satisfaction improves yet another element of the Quadruple Aim, the patient experience.

Our study sought unit-level sick time data from the organizations, but we were sometimes unable to access that data, and we also encountered varied definitions of sick time. In future replication studies, we recommend the use of self-report of sick time since the correlation between self-reports of sick time and sick time recorded by the organization was .80 ($n = 184$, $p = < .01$), meaning that self-reports matched what the organization reported 80% of the time (Gaudine & Gregory, 2010). If a question was added to the survey for nurses to report how many sick days they reported in the previous 4 weeks, that may provide more data to add to the strength of the proposed relationship between a good work experience and decreased sick time.

Recommendations Based on Findings

The next phase of this research trajectory would be to study turnover and sick time at the individual nurse level, to increase the sample size, and to add patient outcomes such as falls, pressure ulcers, and hospital-acquired infections. As caring, clarity, and job satisfaction improve, we would presumably experience an improved healthcare system.

As we found evidence that a positive experience of work for nurses is a direct predictor of both reduced costs and improved patient experience, the authors of this chapter believe that testing more variables related to the Quadruple Aim is a worthwhile endeavor. The sooner we can learn about what specific variables can improve health systems, the sooner we can make operational changes that improve the system.

This research study provides the opportunity to fill a gap in the literature by testing a multivariate model of the experience of work that can be used to more deeply test the Quadruple Aim. Literature on the Quadruple Aim that examined how the experience of staff related to improved patient outcomes was difficult to find. In fact, most of the literature on the Quadruple Aim did not include any research, but simply reported the importance of connecting the experience and satisfaction of the workforce to patient outcomes (Batcheller, Zimmermann, Pappas, & Adams, 2017; Bogetz & Friebert, 2017; Boller, 2017; Bosserman, 2016; Melnick & Powsner, 2016; Sikka, Morath, & Leape, 2015; Valentine, 2018). In a recent systematic review of the literature on the Quadruple Aim, no literature was found that demonstrated the relationship between physician burnout (an important aspect of the overall work experience) and patient outcomes (Rathert, Williams, & Linhart, 2018); however, Brown-Johnson et al. (2019) did find a relationship between clinician satisfaction with the work environment and the patient's likelihood to recommend the hospital.

Examination of the literature (outside of Quadruple Aim literature) which studied the relationship between clinician satisfaction with the work environment and patient outcomes was inconclusive.

Use of literature from within and outside of the Quadruple Aim literature, combined with a multivariate theoretical framework of the experience of work like what has been tested here, may provide a scientifically sound method for advancing the Quadruple Aim—particularly to study the experience of work as a predictor of an improved healthcare system. In addition, using a multivariate model like we have proposed here can provide specific information on four constructs from the work experience—clarity of role, clarity of system, caring for self, and caring of the unit manager—that can be addressed by organizational leaders for the purpose of operational refinement.

As you will recall from Chapter 6, findings from this study that staff members who scored well on the Profile of Caring®—meaning they had high scores in clarity of role, clarity of system, caring for self, the caring of the unit manager, and job satisfaction—would also have good patient outcomes. However, as we learned in our initial work in Jamaica, nurses who have high job satisfaction can have low clarity, and nurses who have high clarity can still have low job satisfaction. This is why studying all of the variables in the Profile of Caring together in a structural equation model is so important. It is the only way we can consider how they behave *together* as a profile, especially when we are evaluating them within complex contexts, like those in which a new framework of care or a caring science initiative is also in effect.

You will recall in Chapter 15 that nurses evolved internally as they became clearer in their roles as the Relationship-Based Care® (RBC) framework was implemented on their units. Thus, it is important that the covariance of all constructs in the Profile of Caring are monitored continually, so we can better understand how the evolving beliefs and behaviors of the healthcare team relate independently and together in relationship to all of the improved health outcomes we're pursuing as the framework of care is implemented.

There are so many things the Profile of Caring can help us discover: At what stage in the implementation of a framework of care does the Profile of Caring for the care team "peak" in relationship to safety for the patient? How does the nurses' self-care impact safety, and can monitoring over time help us detect when lack of self-care converts the Profile of Caring into a profile of burnout? These are questions yet to be answered, but this initial study provides affirmation that the Profile of Caring is indeed a valid set of variables internationally for nurses.

The Profile of Caring may even help bring us closer to achieving the Quadruple Aim itself. Will the study of the Profile of Caring show that it is the same healthcare workers who experience clarity of role, clarity of system, caring for self, the caring of their unit manager, and job satisfaction who are the very people needed to bring healthcare closer to achieving the Quadruple Aim?

17

Developing a Customized Instrument to Measure Caring and Quality in Western Scotland

Theresa Williamson, Susan Smith, Jacqueline Brown, and John W. Nelson

One of the privileges of working outside of one's home country is the opportunity to learn firsthand about cultural differences. As the two previous chapters have reported, there are aspects of nursing practice that appear to be common to nurses everywhere, and there are aspects of nursing practice—and of healthcare in general, for that matter—that are unique to certain cultural contexts.

There are aspects of nursing practice that are unique to certain cultural contexts.

In the international study outlined in Chapter 16, measuring *caring* was complicated by the fact that the tool to measure caring, which proved valid for all of the other contexts studied, did not prove valid for Western Scotland. Because the researchers were not able to use the same instrument for Scotland, some of the data we collected for our colleagues there was not included in the larger international study. It was, however, put to excellent use by our Scottish colleagues.

This chapter reports on the process we went through to create an instrument to measure caring in Western Scotland which could then be further customized to measure the care experience of both patients and nurses. It then shares the ways in which we used our new valid instrument to discover important information about what contributes to a positive work experience for nurses in this unique environment. It is an excellent example of what it takes to make sure you are working with theories that are meaningful and relevant to the people whose experience you are attempting to measure.

Using Predictive Analytics to Improve Healthcare Outcomes, First Edition.
Edited by John W. Nelson, Jayne Felgen, and Mary Ann Hozak.
© 2021 John Wiley & Sons, Inc. Published 2021 by John Wiley & Sons, Inc.

Developing an Instrument to Measure Caring as Perceived by the Patient

We decided to develop an instrument to measure caring for use in the patient population first, as their experience of care was of primary concern and there was a group of volunteers already routinely collecting data from patients for the organization. Once the instrument to measure caring as perceived by patients was developed, the instrument could then be used to create instruments to measure caring as perceived by the staff, including staff's report of caring for the patient, caring for self, and staff's perception of how caring their senior charge nurses (the term used in Scotland for unit manager) were.

In 2013, data scientist Dr. John W. Nelson was invited to the Golden Jubilee, a National Hospital Board, for NHS Scotland in Western Scotland to conduct research to monitor the success of the initial implementation and the eventual sustainment of a framework of caring called the Caring Behaviors Assurance System© (CBAS), which is described in detail in Chapter 18. The first step in measuring a framework of caring is figuring out what could be measured that would give an accurate picture of the framework's success or failure. Caring was the first priority, as caring was what proponents of the CBAS model assured us would be practiced by staff and perceived by patients when CBAS was successfully implemented.

Prior to this work, the 10 Caritas processes of caring in Watson's Theory of Transpersonal Caring (2008a), as measured by the psychometrically sound Caring Factor Survey (CFS), had been found to be a valid construct of caring as perceived by patients in multiple countries (Nelson & Watson, 2019). Thus, we decided to use the CFS to measure changes in the process of caring throughout the implementation of CBAS. Exact wording of the original 10 items in the CFS follows. The headings have been added to indicate at a glance what sort of information was being sought.

Watson's Caring Factor Survey (CFS)

Loving Kindness
1) Every day I am here, I see that the care is provided with loving kindness.

Decision Making
2) As a team, my caregivers are good at creative problem solving to meet my individual needs and requests.

Instill Faith and Hope

3) The care providers honored my own faith, helped instill hope, and respected my belief system as part of my care.

Teaching and Learning

4) When my caregivers teach me something new, they teach me in a way that I can understand.

Spiritual Beliefs/Practices

5) My caregivers encouraged me to practice my own individual spiritual beliefs as part of my self-caring and healing.

Holistic Care

6) My caregivers have responded to me as a whole person, helping to take care of all my needs and concerns.

Helping/Trusting Relationship

7) My caregivers have established a helping-trusting relationship with me during my time here.

Healing Environment

8) This facility and its care providers have created an environment which helps me to heal physically and spiritually.

Promote Expression of Feelings

9) I feel like I can talk openly and honestly about what I'm thinking, because those who are caring for me embrace my feelings, no matter what my feelings are.

Miracles

10) My caregivers are accepting and supportive of my beliefs regarding a higher power which allows for possibilities of healing to take place for me and my family.

A slightly amended version of the CFS was distributed to 446 patients in Golden Jubilee. In this new version, which we called the Healing Compassion Assessment (HCA), Caritas process three was reworded to be less personal and intimate. Caritas process six, regarding holistic care, was divided into two items, and Caritas process eight, regarding the effect of the environment of care, was divided into three statements. Caritas process 10, regarding support in the belief of a higher power, was omitted since it was deemed by the Scottish researchers to not be a valid item for demonstrating caring in Scotland. Wording in several of the items was changed to fit the culture as well. The 12 survey items of the Healing Compassion Assessment follow. Again, the headings are added to indicate what sort of information was sought.

Healing Compassion Assessment (the CFS Survey Customized for Use in Western Scotland)

Practice Loving Kindness
1) My caregivers treat me with kindness and compassion.

Decision Making
2) My caregivers work together to meet my personal needs/requests.

Instill Faith and Hope
3) My requests for information are treated and responded to with respect by my caregivers.

Teaching and Learning
4) When my caregivers talk to me about my care, condition or treatment, they talk in a way that I can understand.

Spiritual Beliefs/Practices
5) My caregivers encourage me to practice my own individual spiritual beliefs whilst in their care. (Do not answer this if you feel spiritual beliefs are too personal.)

Holistic Care
6) My caregivers are knowledgeable
7) My caregivers are skillful

Helping/Trusting Relationship
8) My caregivers have established a helping and trusting relationship with me during my time here.

Healing Environment
9) The environment around me helps me feel better physically.
10) The environment around me helps me feel better mentally.
11) The environment around me helps me feel better spiritually.

Promote Expression of Feelings
12) My caregivers value my feelings (whatever they are) so I can talk openly and honestly about what I am thinking.

Miracles
(omit)

Once we received the results of the patient survey, we used exploratory factor analysis (EFA) to examine whether the newly customized version of the CFS was a valid construct of caring to study within the context of Western Scotland. Methods in the EFA included using Eigenvalues greater than

1.0, principal axis factoring for extraction, and direct oblimin for rotation. Kaiser–Meyer–Olkin (KMO) was used to assess for sampling adequacy and as an indicator of model fit.

A total of 383 of the 446 patients (85.9%) responded to every item of the 12-item Healing Compassion Assessment (HCA) survey, and when factor analysis was then used to examine whether the instrument was valid for the context of Scotland, it showed that Caritas processes 3 and 8 did not load. Since we had already eliminated the item referring to belief in a higher power, this meant that only 7 of the 10 Caritas processes were appropriate to measure caring in this hospital in Western Scotland. An even larger concern, however, was the reports from the data collectors that the patients who were responding to the survey still did not like some of the questions, as they felt the items were too intimate and spiritual in nature. The factor loadings for the 12-item HCA are found in Appendix V.

Because the HCA was not performing well psychometrically, and because patients complained that the survey was too spiritual and intimate, we decided to see whether Swanson's Theory of Caring (2008) would be more appropriate for use in Scotland. An 18-item instrument called the Caring Professional Scale (CPS) measures caring as proposed by Swanson (2008). The CPS had been tested in the United States and was found to be psychometrically sound; it is described in detail in Appendix K of this book.

The same methods for exploratory factor analysis were used to study 15 of the 18 items of the CPS, to see whether they comprised a valid instrument for measuring caring in the context of Western Scotland. The three items we omitted were negative characteristics such as the clinician being emotionally distant, abrupt, and insulting. These three characteristics were determined to occur rarely if ever, and thus were omitted. Analysis of results from 582 patients revealed that 12 of the 15 remaining items formed a 2-factor solution (meaning items arranged themselves into 2 distinct groups). Unfortunately, three of the items did not load, and four items cross-loaded, which means the item could load just as well into either of the two factors of the CPS. Both the failure to load and cross-loading suggested that this 15-item measure may not be valid for Western Scotland, especially considering that the sample of patients was over 400, which is adequate for validating an instrument using factor analysis (Tabachnick & Fidell, 2007). More noteworthy, and similar to our experience with the CFS, was the patients' report to those collecting the data that some of the statements were too much about feelings and were not relevant to what was perceived to be caring in the care provider–patient relationship. Results of the factor loadings for the 15 items of the CPS are noted in Appendix W.

After working with established instruments based on two theories (Watson's and Swanson's) which were found to have some valid and some invalid items for the context of Western Scotland, we decided to use the items from Swanson's CPS instrument that had performed well and round out the rest of the survey with items that aligned with a theory of caring proposed in 2010 by the National Health Service (NHS) in Scotland (The Scottish Government, 2010). In a report entitled The Healthcare Quality Strategy for NHSScotland [sic], the NHS identified what have become known as the "7 Cs" that contribute to quality healthcare. The 7 Cs are Care, Compassion, Communication, Collaboration, a Clean Environment, Continuity of Care, and Clinical Excellence. The Caring Behaviors Assurance System (CBAS), which is based on the 7 Cs, was already being implemented in the organization, so it made sense to develop a new instrument that aligned with the 7 Cs.

We found that there were already nine items from the CPS that aligned with four of the 7 Cs. To capture the rest of the 7 Cs and to add the dimension of measuring quality while we measured caring, quality items already in use within the hospital were added to the instrument. The instrument we eventually used, which combined the CPS items deemed valid in Western Scotland, the items to measure the 7 Cs, and several additional quality items was called the Healing Compassion Survey—7 Cs NHS Scotland (Patient/Family Version).

Note: in the actual survey there are no headings and the order of the questions differed.

The Healing Compassion Survey – 7 Cs NHS Scotland (Patient/Family Version)

Caring

1) Did you feel that the member of staff who just looked after you was comforting?
2) Did you feel that the member of staff who just looked after you was caring?

Compassion

3) Did you feel that the member of staff who just looked after you was understanding?
4) Did you feel that the member of staff who just looked after you was personal (treated you kindly and as an individual)?
5) Did you feel that the member of staff who just looked after you was supportive?

Communication

6) Did you feel that the member of staff who just looked after you was informative?
7) Did you feel that the member of staff who just looked after you was an attentive listener?

Clinical Excellence

8) Do you feel that the member of staff who just looked after you was clinically competent?
9) Did you feel that the member of staff who just looked after you was technically skilled?

Collaboration

10) Did you feel the member of staff who just looked after you worked effectively with others in teams?

Clean Environment

11) Did the member of staff who just looked after you make sure you had everything you needed in your environment of care?

Continuity of Care

12) Did you tend to have the same members of staff take care of you throughout your stay here, or did you have many different members of staff throughout your stay here?

Additional Quality Items

13) Did you feel the member of staff who just looked after you took responsibility to do their job well?

14) Did the member of staff who just looked after you demonstrate their commitment to quality?
15) Did you feel that the member of staff who just looked after you was respectful of you?
16) Did you feel that the member of staff who just looked after you displayed a "can do" attitude at every opportunity?
17) Did the member of staff who just looked after you help you be as involved in your care as you wanted to be?
18) Did you feel that the member of staff who just looked after you address your faith and belief needs as appropriate to their role?

Our new survey was tested on 1,215 patients or family members who responded to all 18 items of the Healing Compassion Survey—7 Cs NHS Scotland. With their answers, we conducted an exploratory factor analysis (EFA), using the same procedure we used for the CFS and CPS. Results revealed three factors (meaning items arranged themselves into three distinct groups) with spiritual care loading all alone in the third factor. Spiritual care had a very low factor loading of .110, but it also cross-loaded with the second factor where the loading was .330. This three-factor structure had good model fit as evidenced by a Kiser–Meyer–Olkin score of .942, but it was undesirable to have a single-item third factor and thus a second EFA was conducted.

In the second EFA, the 18 items were examined again, but instead of using Eigenvalues greater than 1.0 to discover how the items would organically load into separate factors, the 18 items were forced into 2 factors. Results of this second EFA revealed that all 18 items loaded, demonstrating that it was a valid instrument for measuring caring in Western Scotland. Only the item about spiritual care had a factor loading below .3 (.266) which suggested it may not belong to this construct of caring in Western Scotland. Table 17.1 shows the survey items ranked by factor loading. You can see in the columns on the right that the first 14 items comprise one factor and the last 4 items comprise a second factor.

These factor loading scores were deemed sufficient for this tool to become the instrument with which we would measure caring as perceived by the patient and/or family in Western Scotland.

Table 17.1 Factor loadings for the healing compassion survey—7 Cs NHS Scotland (patient/family version).

	Loading in rank order	
Caring/quality items	1	2
Q15: Did you feel the member of staff who just looked after you took responsibility to do their job well?	.923	
Q16: Did the member of staff who just looked after you demonstrate their commitment to quality?	.864	
Q8: Did you feel that the member of staff who just looked after you was supportive?	.848	
Q7: Did you feel that the member of staff who just looked after you was caring?	.837	
Q11: Did you feel that the member of staff who just looked after you was respectful of you?	.819	
Q5: Did you feel that the member of staff who just looked after you was understanding?	.813	
Q6: Did you feel that the member of staff who just looked after you was personal (treated you kindly and as an individual)?	.803	
Q17: Did you feel the member of staff who just looked after you worked effectively with others in teams?	.675	
Q9: Did you feel that the member of staff who just looked after you was an attentive listener?	.653	
Q2: Did you feel that the member of staff who just looked after you was comforting?	.650	
Q18: Did you feel that the member of staff who just looked after you displayed a "can do" attitude at every opportunity?	.585	
Q4: Do you feel that the member of staff who just looked after you was clinically competent?	.426	
Q3: Did you feel that the member of staff who just looked after you was informative?	.402	
Q10: Did you feel that the member of staff who just looked after you was technically skilled?	.321	
Q13: Did the member of staff who just looked after you make sure you had everything you needed in your environment of care?		.463
Q14: Did the member of staff who just looked after you help you be as involved in your care/treatment as you wanted to be?		.430
Q19: Was the member of staff who just looked after you consistently part of the team who looked after you throughout your stay here?		.307
Q12: Did you feel that the member of staff who just looked after you was able to address your faith and belief needs as appropriate to their role?		.266

Developing an Instrument to Measure Caring as Perceived by the Nursing Staff

An instrument to assess the process of caring as perceived by nurses was developed to mirror the instrument used for patients. In this version, the wording was revised to inquire about care from the provider's perspective.

Again, using Eigenvalues greater than 1.0 and the same methods of extraction and rotation, the factor analysis revealed that the 18 items comprised one single construct of caring and quality with 3 factors: one for caring, one for the process of care, and one for quality. The instrument was named the Healing Compassion Survey—7 Cs NHS Scotland (Staff Version); the 18 items in the survey are as follows.

The Healing Compassion Survey—7 Cs NHS Scotland (Staff Version)

1) For the patients you took care of, were you supportive?
2) For the patients you took care of, were you caring?
3) For the patients you took care of, did you demonstrate your commitment to quality?
4) For the patients you took care of, were you an attentive listener?
5) For the patients you took care of, were you informative?
6) For the patients you took care of, did you display a "can do" attitude at every opportunity?
7) For the patients you took care of, did you make sure they had everything they needed in their environment of care?
8) For the patients you took care of, were you understanding?
9) For the patients you took care of, did you take responsibility to do the job well?
10) For the patients you took care of, were you personal (did you treat the patient kindly and as an individual)?
11) For the patients you took care of, were you respectful of the patient?
12) For the patients you took care of, were you technically skilled?
13) For the patients you took care of, were you able to encourage the patients to make decisions about their care/treatment?

14) For the patients you took care of, were you clinically competent?

15) For the patients you took care of, did you work effectively with others in teams?

16) For the patients you took care of, were you comforting?

17) For the patients you took care of, were you able to ensure that their faith or belief needs were met?

18) For the patients you took care of, have they consistently been part of your patient assignment?

The instrument was administered to 436 nurses on wards on which CBAS was being implemented. The order of the questions, as presented here, is not the order in which they were presented to nurses; they are presented here in order of those having the highest factor loading (.750) to the lowest factor loading (.376). A factor loading reveals the strength of the relationship of the survey item with the variable of interest—in other words, it shows how valid the item is to measure what you say you want to measure. You can see the factor analysis and validity ranking of all 18 items for staff members' perception of their care of patients in Appendix X.

Building a Structural Model for Assessing How Caring, Clarity, and Job Satisfaction Relate to One Another in Western Scotland

Although it took us a lot of time and testing to devise valid instruments to measure caring as perceived by both the patients/families and staff members, that was only part of our work. It was important to create an instrument to measure caring that was concise, so that a complete model of the work experience could be examined that included caring for self, caring of the senior charge nurse (ward manager), clarity of role and system, and 11 dimensions of job satisfaction. Having a multivariate model to study the experience of nurses would facilitate planning for how to improve nurses' experience of work by understanding what impacts their job satisfaction. Figure 17.1 provides a structural model indicating what we sought to study.

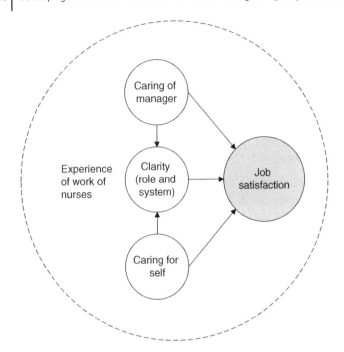

Figure 17.1   Structural model to study the experience of work of nurses in Western Scotland.

To develop an instrument to measure nurses' overall experience of work, we used items from a variety of sources. Job satisfaction was already being measured by the Healthcare Environment Survey (HES). The Scottish version of the HES is a 56-item instrument used to assess 11 factors of social and technical aspects of the work of nurses (Nelson & Cavanagh, 2017). To assess clarity, the 15-item instrument developed by Felgen and Nelson (2016) and described in Chapter 15 was used. There were also 10 demographic items in the survey, bringing the sum of items for the HES, clarity instrument, and demographics to 82 items. Due to the high number of items staff members were going to be asked to respond to, we wanted to measure caring for self and caring of the senior charge nurse as perceived by staff using as few items as possible. Thus, instead of using all 18 of the items we found valid to measure caring in Western Scotland, only 7 items from the Healing Compassion Survey—7 Cs NHS Scotland (Staff Version) were selected to measure caring for self and the caring of the senior charge manager. The seven selected items covered the first three C's of the 2010 NHS report: (a) Care (caring and comforting), (b) Compassion (supportive, understanding, and personal), and (c) Communication (attentive listener, informative). These seven items were customized twice—once to develop seven items to measure caring for self and again to develop seven more items to measure the caring of the senior charge nurse. (Factor loading and ranking of these 14 items can be found in Appendix Y.) This brought the total number of items to measure the experience of nurses to 96 items. There were also 3 qualitative items and 1 global measure of job satisfaction, for a total of 100 items.

Results

This section will review results of the most recent study of the patient's report of caring and experience of work of nurses in Western Scotland, comparing scores over time when possible.

Results: Caring and Quality, as Reported by the Patient

Data was derived from 17 patient care wards over four quarters. Prior to the Caring Factor Survey proving to be invalid for Western Scotland, it was used as our patient survey for four quarters. The sample sizes for these quarters were 27, 230, 189, and 82. Mean scores, as noted in Figure 17.2, reveal that caring

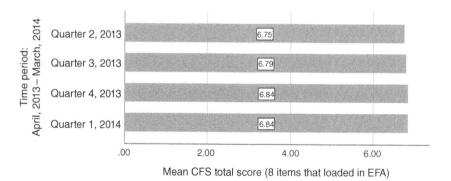

Figure 17.2 Caring as reported by patients over four consecutive quarters.

scores were all over 6.7 on a 7-point Likert scale, which indicates a very high level of patients' report of caring. The mean score improved slightly over time, but this increase was not statistically significant using an alpha of .05.

The validated Healing Compassion Survey—7 Cs NHS Scotland (Patient/Family Version) was given to 2,941 patients and/or family members over the course of 60 months. Prior to analysis of the survey, the faith item was removed due to low factor loading from the exploratory factor analysis as described earlier in this chapter. The factor analysis was conducted when 1,825 patients had responded to all 17 remaining items in the Healing Compassion Survey—7 Cs NHS Scotland (Patient/ Family Version). With this new instrument, mean scores were all above 4.8 on a 5-point Likert scale, indicating, again, a high level of patients' report of caring and quality (see Figure 17.3).

It is worth noting that although the CFS was eventually deemed invalid for Western Scotland, the scores received using the CFS and the Healing Compassion Survey—7 Cs NHS Scotland (Patient/Family Version) were virtually identical.

Table 17.2 contains more in-depth data from this same round of surveys. Note that there is almost no variance in the patients' report of caring, as is noted by the very small standard deviation. This indicates that there was high consistency among the

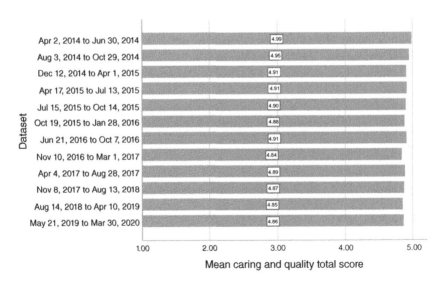

Figure 17.3 Caring as reported by patients over 73 months (2014–2020).

Table 17.2 Patients' report of caring, April 2, 2014 to March 30, 2020.

Date for each dataset	N	Mean	Std. deviation	Minimum	Maximum
April 2, 2014 to June 30, 2014	206	4.99	.06	4.47	5.00
August 3, 2014 to October 29, 2014	218	4.95	.26	2.65	5.00
December 12, 2014 to April 1, 2015	238	4.91	.17	2.94	5.00
April 17, 2015 to July 13, 2015	218	4.91	.12	4.06	5.00
July 15, 2015 to October 14, 2015	238	4.90	.21	2.12	5.00
October 19, 2015 to January 28, 2016	234	4.88	.17	3.41	5.00
June 21, 2016 to October 7, 2016	239	4.91	.08	4.41	5.00
November 10, 2016 to March 1, 2017	237	4.84	.27	2.35	5.00
April 4, 2017 to August 28, 2017	221	4.89	.12	3.59	5.00
November 8, 2017 to August 13, 2018	205	4.87	.24	1.88	5.00
August 14, 2018 to April 10, 2019	230	4.85	.15	3.53	5.00
May 21, 2019 to March 30, 2020	191	4.86	.12	4.18	5.00
Total	2675	4.90	.18	1.88	5.00

patients about what it meant to demonstrate caring behaviors. The consistently high mean scores denote that patients felt extraordinarily well cared for by their nurses for the entire 6 years of testing.

Results: Job Satisfaction as Reported by the Nursing Staff

Using a 7-point Likert scale with higher scores indicating greater job satisfaction, mean scores for each of the 11 dimensions of job satisfaction ranged from 3.75 (satisfaction with distributive justice) to 5.80 (satisfaction with patient care). A midpoint of 4.0 indicated neutrality of respondents neither agreeing nor disagreeing that they were satisfied with their jobs. Standard deviations for all individual facets were above 1.0, indicating some variance in how individuals perceive the various facets of their jobs. High scores, low scores, mean scores, and standard deviations for all constructs and facets of constructs of nurse job satisfaction for the final survey year are reported in Table 17.3.

You can see, closer to the bottom of Table 17.3, that constructs of clarity and caring, which were also measured using a 7-point Likert scale, all had mean scores above 4.0. Higher scores indicate higher levels of clarity and caring. Caring had standard deviations above 1.0, while clarity had standard deviations below 1.0, indicating that respondents had a more varied experience of caring for themselves and the care of the senior charge nurse, while respondents had a somewhat more uniform experience of clarity of self, role, and system.

Comparing Job Satisfaction Over Time

Job satisfaction was studied by aggregate and individual ward over time. Results revealed that there was no significant change at the aggregate level of job satisfaction during the 6-year period we studied, with mean scores of 5.06 (Study 1), 5.06 (Study 2), and 5.04 (Study 3).

However, a general linear model (GLM) analysis revealed that some individual wards *did* have a change in job satisfaction over time. In fact, seven wards improved their job satisfaction mean score between the second and third measures. Changes in scores for all participating wards are noted in Figure 17.4.

Table 17.3 Descriptive statistics for all constructs and facets of constructs of nurse job satisfaction, final survey year only.

Constructs and facets of constructs	N	Lowest received score	Highest received score	Mean	Standard. deviation
Satisfaction with relationship with coworker (5 item)	399	1.60	7.00	5.46	1.16
Satisfaction with relationship with RN	399	1.00	7.00	5.33	1.19
Satisfaction with relationship with coworker (10 item)	396	1.40	7.00	5.39	1.12
Satisfaction with relationship with doctor	394	1.20	7.00	4.86	1.30
Satisfaction with workload	392	1.00	7.00	4.99	1.30
Satisfaction with autonomy	391	1.00	7.00	5.13	1.36
Satisfaction with distributive justice	385	1.00	7.00	3.72	1.58
Satisfaction with patient care	374	1.00	7.00	5.80	1.02
Satisfaction with participative management	378	1.00	7.00	5.25	1.61
Satisfaction with professional growth	380	1.00	7.00	4.91	1.59
Satisfaction with executive leadership	373	1.00	7.00	4.66	1.39
Satisfaction with scheduling	378	1.00	7.00	5.13	1.52
Satisfaction with resources	379	1.00	7.00	5.48	1.19
HES Total (job satisfaction)	356	2.02	7.00	5.07	.98
Clarity of role	370	1.40	7.00	5.80	.92
Clarity of system	362	1.40	7.00	5.68	.94
Clarity (role and system)	358	1.40	7.00	5.72	.87
Caring for self	348	1.00	7.00	5.20	1.14
Caring of manager (senior charge nurse)	357	1.00	7.00	5.51	1.50

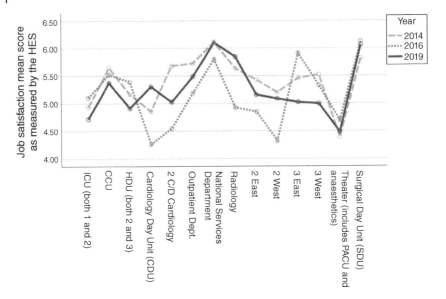

Figure 17.4 Interaction of time and ward for mean scores of job satisfaction.

Testing the Final Model to Measure the Experience of Nurses

Because we found a change in job satisfaction over time for individual units, we wanted to understand what might be driving the change. Thus, the model of the experience of work (Figure 17.1) was studied further, using regression analysis. A post hoc power analysis, using software G-Power, revealed that an alpha of .001 would be needed for a sample size this large. The power was .988, the alpha was .001, and the effect size was .15. Results revealed that the only statistically significant predictor of job satisfaction was the ward the respondent worked on. Results of all the individual regression equations related to job satisfaction and ward are noted in Appendix Z.

Testing of the final model also revealed that clarity of self, role, and/or system was a significant predictor of job satisfaction, so we examined demographics in relationship to clarity. Again, the ward the person worked on was the only predictor of whether staff members were experiencing clarity of self, role, and/or system. Demographics and their relationship to the dependent variable of clarity are noted in Appendix AA.

Using hierarchical regressions, after three surveys were completed, it was found that the ward the person worked on, the level of clarity the person had about self, role, and system, and the caring of the senior charge nurse were all significant predictors of nurse job satisfaction.

The ward the person worked on, the level of clarity the person had about self, role, and system, and the caring of the senior charge nurse were all significant predictors of nurse job satisfaction.

Ward predicted 20.1% of the variance of job satisfaction ($p = <.001$), clarity of self, role, and/or system predicted 42.3% of the variance ($p = <.001$), and the caring of the senior charge nurse predicted 8.1%. In total, these three variables predicted 70.6% of the variance of job satisfaction of this sample. Caring for self was not found to have a statistically significant relationship to job satisfaction.

These variables are the same as those discussed in previous chapters as they relate to the Profile of Caring®, which measures the degree to which nurses are experiencing clarity of role, clarity of system, caring for self, caring of manager, and job satisfaction. It was affirming to find that despite the difference between caring theory used in Scotland and the other seven countries in our international study, the result was the same: The five variables in the Profile of Caring are appropriate for use in Western Scotland, too. This is particularly beneficial for the organizations in Scotland implementing CBAS, as the Profile of Caring can isolate and measure the experience of nurses, even in complex contexts such as those in which frameworks of care delivery and caring science theories are being implemented.

Due to clarity having a significant relationship with job satisfaction, and its being an important variable in the original hypothesized model, we wanted to know what predicted clarity. Hierarchical regression was again used to examine clarity. Among the demographics tested, only the ward the person worked on was found to be a statistically significant predictor of clarity.

Three models were tested. Using the model that initially measured how clarity related to job satisfaction, we made three more models, adding ward for the first model, caring of the senior charge nurse for the second model, and caring for self for the third model. Figure 17.5 shows the third model.

Results from testing Model 3 revealed that ward predicted 13.7% of the variance of clarity ($p = <.001$), caring of the senior charge nurse predicted 21.1% of the variance of clarity ($p = <.001$), and caring for self predicted 6.8% of the variance of clarity ($p = <.001$). Combined, these three independent variables predicted 41.6% of clarity.

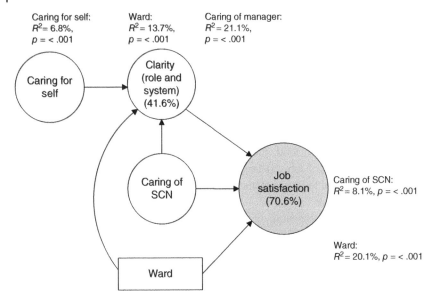

Figure 17.5 Model 3: testing job satisfaction, caring, and clarity.

Imagine how valuable this information was to the team in Scotland! Knowing so precisely what predicted job satisfaction, the team could take well-targeted action to improve it.

Discussion

The team at this organization was left with something that will provide value for years to come: two versions of a fully vetted instrument—the Healing Compassion Survey—7 Cs NHS Scotland—to measure caring and quality in Western Scotland, which specifically aligns with the caring values proposed by the National Health System of Scotland. This instrument was essential to the organization's ability to not only create change, but to monitor and sustain the kind of change that leads to cultural transformation.

Going forward, the Healing Compassion Survey—7 Cs NHS Scotland can be used to identify, for example, which staff members are fully engaged and which staff members may need additional support to feel like they are part of the team and/or fully engaged members of the organization. Or, if someone in the organization has a hunch about what impacts or is impacted by acts of caring, a model like the one in Figure 17.5 can now easily be adapted to test the hunch, and the findings can be used to improve practice. For example, if it is proposed that a class on active listening (looking for verbal and nonverbal cues and responding to them) will increase perception of caring by the one being listened to, the Healing Compassion Survey—7 Cs Scotland could be used as the core of a pre- and post-class study. Since this organization now also has a validated instrument to measure caring of the managers, this same practice of active listening could be evaluated by staff members to measure how their relationships with their managers are affected by the practice of active listening. The possibility to test "hunches" and formal theories is vast when instruments are available that are designed and validated for your specific context. With measurement instruments validated rigorously for context, we can connect behavior and belief to safety and cost.

With measurement instruments validated rigorously for context, we can connect behavior and belief to safety and cost.

Chapter 18 continues the story of our work in Scotland with a report of how the work of developing an appropriate way to measure caring and discovering what predicted job satisfaction in Scotland contributed to this group's ability to measure the success of its implementation of the Caring Behaviors Assurance System (CBAS) care delivery model.

18

Measuring the Effectiveness of a Care Delivery Model in Western Scotland

Theresa Williamson, Susan Smith, Jacqueline Brown, and John W. Nelson

In 2010 the NHSScotland [sic] published a report called "The Healthcare Quality Strategy for NHSScotland," which introduced the seven behaviors of caring, or "the 7 Cs": Care, Compassion, Communication, Collaboration, a Clean Environment, Continuity of Care, and Clinical Excellence (The Scottish Government, 2010). Two years later, it made person-centered care one of its strategic priorities. To meet this new requirement, leadership in Golden Jubilee, a National Hospital Board, for NHS Scotland in Western Scotland, along with 17 other hospitals, proposed implementing the Caring Behaviors Assurance System© (CBAS), a person-centered care delivery system specifically designed to operationalize the seven behaviors of caring listed in the 2010 NHS report.

To measure the overall effectiveness of CBAS throughout its implementation, several instruments were used to assess the extent to which constructs taught in CBAS, as well as the larger system that supported CBAS operations, were being accomplished in practice. Staff members and leaders worked with data analyst, Dr. John W. Nelson, to understand the data and the associated and perceived impacts of CBAS, and to build models of measurement that would capture the changes made over the 7-year implementation. Measurement was taken at the start of implementation and again every 24 months over a 7-year period. This chapter reports on the instruments used to assess the effectiveness of CBAS, and it concludes with a review of how the resulting data was used to understand what operational changes

Using Predictive Analytics to Improve Healthcare Outcomes, First Edition.
Edited by John W. Nelson, Jayne Felgen, and Mary Ann Hozak.
© 2021 John Wiley & Sons, Inc. Published 2021 by John Wiley & Sons, Inc.

were most likely to improve outcomes and how those changes eventually did improve outcomes at the ward level.

The Caring Behaviors Assurance System (CBAS)

CBAS is based on the belief that it is incumbent upon those delivering care to establish common understanding with patients and families, and to take action when care falls below the organization's standards. A consulting firm called Choice Dynamic International is the creator of the model, which was commissioned by NHSScotland. Implementation in the organization referenced in this chapter was led by Dr. Susan Smith and Janina Sweetenham.

According to Choice Dynamic International, CBAS is designed to (a) raise quality standards, (b) create employee accountability, (c) make self-care a priority for clinicians, and (d) foster a deep sense of employee ownership for the outcomes of care delivery. The aims of CBAS are amplified here:

Raise Quality Standards

The educational components of CBAS, along with the structure of its implementation, uncover the perceptions of staff members regarding care and compassion so operational changes can be put into place that help all care to be consistent with organizational standards.

Create Employee Accountability for Quality and Safety in the Care Experience

Staff members are trained to assess their own compliance with quality and caring behaviors and take ownership for designing and implementing any necessary operational changes. CBAS is an empowering model that encourages staff members in all roles and at all levels to deal directly with issues as they arise and to join others in changing any structures, processes, and policies that are not working.

Establish the Self-Care of Clinicians as Central to Caring

Staff members actively foster their own mental, emotional, and physical health so they can deliver the best care.

Establish Ward-Level Employee "Ownership" for Care Delivery

Responsibility for implementation of CBAS is owned at the ward level. Each ward identifies three to five Quality Champions (nurses, doctors, healthcare assistants, catering staff, porters, allied health professionals), to lead the process of change on the ward. The Quality Champions use an assessment called the Person-centered Care Quality Instrument (PCQI) to assess the extent to which their ward is successfully actualizing the dimensions of the CBAS model. The assessment is conducted every 6 months, and findings from this assessment are used to determine which dimensions of CBAS merit specific focus over the next 6 months on each ward. These assessments also enable staff members to revisit the reason they came to the work of caring in the first place, and they provide an opportunity for staff members at all levels to share their views.

To align the CBAS model with the specific context of Scotland, the 7 Cs of quality care proposed by the National Health Service (NHS) Scotland were integrated into all existing CBAS materials, trainings, and assessments. Again, the 7 Cs are Care, Compassion, Communication, Collaboration, a Clean Environment, Continuity of Care, and Clinical Excellence. As you can see in Figure 18.1, the 7 Cs are the basis for the six dimensions of CBAS-Scotland, which combines care and compassion into one dimension.

Figure 18.1 Caring Behavior Assurance System (CBAS) model—Scotland.

Implementation of CBAS

To prepare themselves to implement CBAS, the Quality Champions attend a 3-day educational program to learn the systems of CBAS and how to use the PCQI assessment tool. A depiction of the entire implementation process is provided in Figure 18.2.

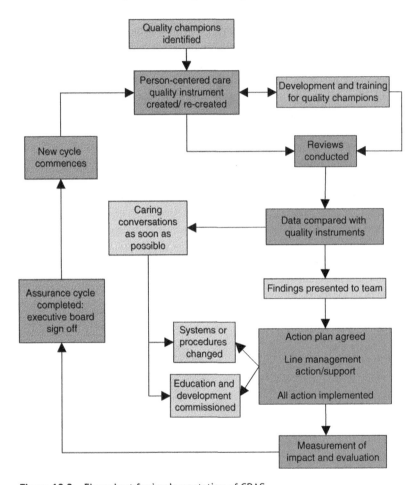

Figure 18.2 Flow chart for implementation of CBAS.

At the organizational level, appropriate people are selected to attend a 5-day training program led by Choice Dynamic International (CDI), with the intention that they will become their organization's in-house CBAS facilitators. The program is designed to prepare them to teach the principles of the CBAS program to their peers and to commence creating their area's individual PCQI. After receiving the training led by CDI, these aspiring facilitators are observed delivering the 3-day training program to clinical teams and/or to others desiring to learn the CBAS program. If they are observed to be skilled facilitators of the material, they are permitted to run the 3-day program independently, as in-house facilitators. Refresher workshops are periodically offered to clinical nurse managers and senior charge nurses to ensure that leadership engagement with the program continues.

The stages of implementation of CBAS in the organizations we studied are described here:

1) In April 2012 3-day trainings commenced for clinical staff members and leaders with a desire to learn more about CBAS. Most attendees were registered and non-registered nurses, and attendees became Quality Champions once training was completed.
2) Mid-2013 the Quality Champions from most wards had completed the CBAS training.
3) By the end of 2013 all departments had sent at least three members of their nursing team to CBAS training and therefore all clinical areas had at least three Quality Champions. From then on, it was up to each senior charge nurse or clinical nurse manager to determine how many staff members to send to subsequent CBAS trainings to become Quality Champions.
4) In 2014 members of the medical staff were introduced to CBAS. (We were committed to orienting the medical staff even though our research did not extend beyond nursing.)
5) In December 2014 we commenced a series of annual review days designed to sustain momentum over time.
6) In 2015 clinical nurse managers attended CBAS refresher workshops.
7) In October 2018 senior charge nurses (ward managers) attended a back-to-the-basics CBAS training, which included principles of CBAS, the PCQI, reporting, and updates of action plans.

8) Throughout this process, staff volunteers worked with the data scientist to collect patient data and report back on patients' views of survey items. These volunteers attended a 1-day session where they learned the principles of CBAS, giving them a better understanding of how the survey items pertained to daily practice.

9) Every May and November that the CBAS cycle ran within all clinical areas, and the PCQI documents were completed to assess the six dimensions of CBAS, we developed new actions, revisited existing actions, and revised our plans as needed.

Measurement of CBAS

One existing instrument and two newly developed instruments were used to assess the degree to which the concepts taught in CBAS were being actualized in practice and/or reinforced throughout the implementation of CBAS. This section will address the measurement of the operations of CBAS—specifically (a) caring enacted by staff, (b) operations that helped staff implement CBAS, and (c) how the 11 dimensions of job satisfaction were impacted by CBAS. Measurement of these operations made the concepts of CBAS visible empirically so they could be addressed operationally.

Measurement of operations made the concepts of CBAS visible empirically so they could be addressed operationally.

The Patient-centred Care Quality Instrument (PCQI)

The Patient-centered Care Quality Instrument (PCQI) is an 85-item audit tool used to collect data on the extent to which the 6 dimensions of the construct of caring proposed in the CBAS framework are being actualized in practice. To match the model, the PCQI is divided into six sections:

- Section A: Care and Compassion (32 items)
- Section B: Communication (14 items)
- Section C: Collaboration (12 items)
- Section D: Clean Environment (10 items)
- Section E: Continuity of Care (8 items)
- Section F: Clinical Excellence (9 items)

Theoretically, if the behaviors listed in the PCQI are operationalized, quality care will be perceived by patients, families, and staff members. This chapter details the evidence gathered using the PCQI at Golden Jubilee over a 4-year period.

Each section of the PCQI comprises survey items that measure the multiple facets of the respective dimensions. Staff members who are collecting data using the PCQI can use any of six types of methods to collect data:

1) Caring Walk (CW)—Trainers of the CBAS system walk through a ward and document observed behaviors consistent with training in CBAS.
2) Practice Observation (PO)—Trainers of CBAS observe clinical practice of staff members to document behaviors consistent with training in CBAS.
3) Patient/Family Interview (PFI)—Trainers of CBAS talk with patient and/or the families of patients to verify behaviors consistent with training in CBAS.
4) Manager Conversations (MC)—Discussions with ward managers reveal dimensions consistent with training in CBAS.
5) Paperwork (PW)—Trainers examine electronic or paper documents for evidence of behaviors consistent with training in CBAS.
6) Other (O)—Evidence may be collected on the effectiveness of CBAS that does not fit any of the other five types of evidence.

Patient care wards or departments trained in the CBAS framework used the PCQI every 6 months to identify specific behaviors they wish to focus on in the coming 6 months.

Table 18.1 features four items from the PCQI for Scotland—specifically from the 12-item section on collaboration.

Table 18.1   Four items from Section B (Collaboration) of the PCQI.

30	At the beginning of a shift, patients are informed of who is who and what will happen during that shift (B1)
31	Information is given in a user-friendly way. This means all written and verbal communication is in the correct language and style to maximize comprehension (B2)
32	Staff members introduce themselves and their role at the beginning of a shift (B5)
33	Patients and family members/visitors are always treated with respect. This is shown by: • Smiling • Using comfortable eye contact • Open body language • An offer of help/assistance • Introducing self and role if not yet done (B7)

If the PCQI revealed that these actions (or others like them) were not consistently actualized on a unit, they may be selected for focused implementation and monitoring over the next six months. After 6 months, the PCQI would be administered again, and a new set of actions (sometimes repeating some of the actions from the previous 6 months) would be selected for focused implementation and monitoring. All 85 items of the PCQI address behaviors deemed essential for CBAS to be effective, but since it is impossible for staff members to focus on 85 things at once, a reasonable number of actions were selected for specific focus for each 6-month period. Observation using the PCQI, using each ward's customized version, would happen throughout the 6-month period, and PCQI submission by each ward was targeted to be May and November of each year.

Items from Section A, Care and Compassion, were most commonly selected for focused implementation and monitoring, which is not surprising since Section A has 32 items, the most of any section. On average, each ward selected 13 or 14 of the 32 items under "Care and Compassion" to enact and assess for each 6-month period. Items from Section B, Communication, had the second most items selected for focused implementation and monitoring across all three time periods, but there were half as many items selected from Section B as from Section A. As you can see from Figure 18.3, items from the dimensions of Communication, Collaboration, Clean Environment, Continuity of Care, and Clinical Excellence were selected for focused implementation and monitoring far less frequently than items from the dimension of Care and Compassion.

The Operations of CBAS Assesssment

In 2017, 5 years into the implementation of CBAS, a national report was published stating that CBAS had failed to be implemented successfully on a majority of the wards it examined, and thus was not found to make a difference on quality indicators for patients or staff members (Duncan, Colver, Stephenson, & Abhyanakar, 2017). Specifically, one of the studies in this report examined 13 patient care wards and found 2 wards that had successfully implemented CBAS, 2 wards with complete failure of

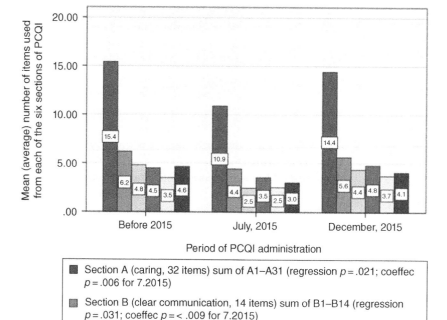

Figure 18.3 Six CBAS dimensions and average number of items used per ward, over time.

implementation, and nine wards with mixed results (Duncan et al., 2017). Unfortunately, the study on which the Duncan report was based used only quantitative data and interviews in a small sample, which suggests that the study may not present a valid argument that CBAS failed. It is also relevant that the people using CBAS every day did not seem to feel, as a rule, that it was ineffective. Dr. Nelson suggested that a truer assessment of how well CBAS was doing would come from examining the two wards that "failed" and the two wards that "succeeded" to understand *why* CBAS appeared to have failed or succeeded. Examination of outliers in this way is referred to as Parato mathematics.

The study by Duncan et al. (2017) identified that staff members having easy access to CBAS Quality Champions contributed to the successful implementation of CBAS. This factor and others noted in the 2017 study, were discussed and affirmed by Dr. Nelson as likely success factors for implementing CBAS. As a result of this examination, a measurement instrument was developed which the team believed could assess and affirm the success factors for CBAS implementation. Use of the assessment throughout the organization would also help strengthen the infrastructure and systems of CBAS in areas reporting low awareness and/or practice of the CBAS dimensions, because regular use of the tool, in and of itself, would improve awareness of CBAS.

Identified success factors were placed into 17 statements and turned into an instrument called the Operations of CBAS Assessment. Staff members would respond to each item using a 7-point Likert scale from strongly disagree (1) to strongly agree (7):

1) I am able to contact CBAS facilitators easily.
2) I know who the CBAS facilitators are for my organization.
3) The CBAS facilitators are always helpful.
4) The CBAS facilitators I work with are expert in helping me understand what behaviors I can change to provide the best patient care possible.
5) I understand how the Person-centered Care Quality Instrument (PCQI) is used to improve patient care.
6) I remember using the PCQI well in my CBAS training.
7) It was clear in our CBAS training how the PCQI could support our team to create action plans for change.
8) I believe CBAS is important for improving the care at this hospital.
9) Coworkers I work with believe CBAS is important for improving the care at this hospital.
10) I feel there is adequate time for me to do what I feel I need to do to help implement our action plan following the CBAS training on my ward/department.
11) There are enough resources for me to feel we will be able to implement our CBAS action plans on my ward/department.
12) The senior charge nurse I work with inspires me to apply what I learned during the CBAS training in patient care.

13) The senior charge nurse is part of the support I feel is help-ful to successfully implement the action plans we developed in the CBAS training.

14) The action plans following the CBAS implementation pro-gram have support from the executive board.

15) Executive board members continue to demonstrate support for the CBAS action plans following the CBAS implemen-tation program.

16) Executive leaders are aware of our team action plans.

17) Executive leaders are supportive of the change agenda that will come from our team action plan.

The assessment was designed to help leaders (a) understand whether the success factors were present, (b) assess whether this was a valid tool to measure operations of CBAS, and (c) cor-relate the operations of CBAS with quality indicators for staff members and patients.

The Healthcare Environment Survey (HES)

The instrument used in the organizations to gather both quan-titative and qualitative data about nurse job satisfaction was the Healthcare Environment Survey (HES), which has been exten-sively tested for validity and reliability throughout the world (Nelson & Cavanagh, 2017; Nelson et al., 2015) and is described numerous times in this book. The dimensions measured on the wards on which CBAS was implemented included four social dimensions (relationships with coworkers, relationships with physicians, relationship with the ward manager, and patient care) and seven technical dimensions (autonomy, workload, distributive justice, executive leadership, professional growth, scheduling, and resources). As with the Operations of CBAS Assessment, all items in the HES used a 1–7 Likert scale from strongly disagree (1) to strongly agree (7).

Findings from the PCQI, the Operations of CBAS Assessment, and the HES

Much was learned about the effectiveness of the CBAS imple-mentation through examining the findings of the Person-centered Care Quality Instrument (PCQI), the Operations of CBAS Assessment, and the Healthcare Environment Survey (HES).

Effectiveness of CBAS Discovered by Using the PCQI

The PCQI was designed to be administered every 6 months so its findings could be used for up-to-the-minute action planning. As mentioned, the dimensions of CBAS were assessed by the Quality Champions through six different methods of assessment (e.g. caring walk, patient/family interview, etc.). The Quality Champions were free to choose their own assessment methods. Early findings indicated the Quality Champions learned perhaps as much about effectively using the assessment methods in the first year as they did about the effectiveness of CBAS itself. Figure 18.4 shows that the Quality Champions were using the assessment method of "other" extensively before they had adequate experience in their role, eventually settling into use of the "caring walk" and "patient/family interviews" as their preferred methods of assessing the 32 different actions of Care and Compassion outlined in Section A of the PCQI.

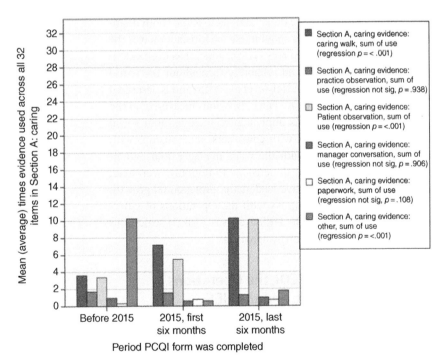

Figure 18.4 Types of evidence used for Section A, care and compassion, 2012–2015.

During the first year of its use, we discovered that while the PCQI was good at measuring what it was intended to measure, it was limited in its ability to truly give a clear picture of how well CBAS was permeating the culture. Findings from all six CBAS dimensions are reviewed here.

Care and Compassion

Review of the data from the PCQI revealed that evidence of Care and Compassion doubled from before 2015 to the second half of 2015. Figure 18.4, which was also used to demonstrate the learning curve of the Quality Champions, shows how incidents of Care and Compassion increased in the first year of CBAS implementation, and the p-value of .006 tells us that this finding would occur by chance only six in 1,000 times.

Communication

Examination of evidence of Communication revealed that eventually the caring walk and patient–family interviews proved the best ways to observe the 14 different actions of Communication outlined in Section B of the PCQI. Again, the dramatic decrease in "other" as a method of observation depicts the time it took for the Quality Champions to learn more about the data collection methods. As Figure 18.5 shows, clear Communication, observed during the caring walk, increased from one to two in the first two time periods and then doubled again from two to four by the third measurement in 2015. From a statistical perspective, this doubling is a big change, as the p-value of <.001 tells us that this finding would occur by chance less than one in a 1,000 times. Statistically significant improvements were observed via the data collection methods of the caring walk ($p = <.001$), patient/family interviews ($p = <.001$), and other methods ($p = <.001$).

Collaboration

When looking for evidence of Collaboration, the Quality Champions again took some time to master the assessment methods. Figure 18.6 indicates that nearly all of the observations of Collaboration were filed under "other" for the first survey period. Fairly quickly, however, they discovered that patient/family interviews proved the best way to observe the 12 different actions of Collaboration outlined in Section C of the

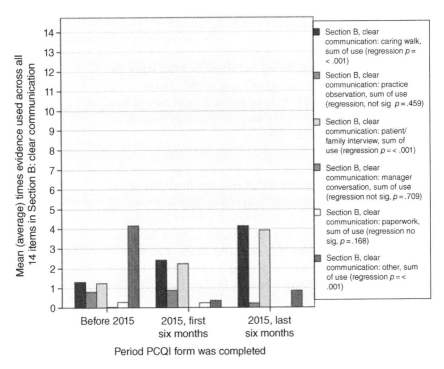

Figure 18.5 Types of evidence used for Section B, communication, 2012–2015.

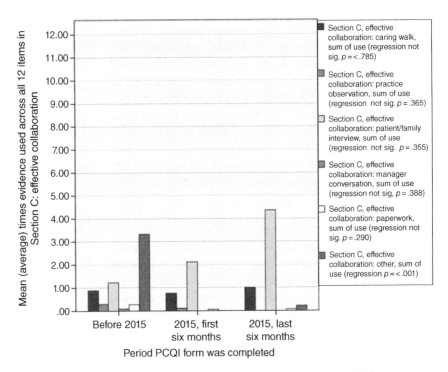

Figure 18.6 Types of evidence used for Section C, collaboration, 2012–2015.

PCQI. You can see in Figure 18.6 that observed and/or reported instances of Collaboration using patient/family interview, quadrupled from measurement one, before 2015, to the second half of 2015. This overall change was not statistically significant, but the quadrupling of evidence using only patient/family interviews is impressive.

Clean Environment

Examination of evidence of Clean Environment revealed that, again, the caring walk and patient–family interviews proved the best way to observe the 10 different actions of Clean Environment outlined in Section D of the PCQI. As Figure 18.7 suggests, while the observation methods changed predictably, actualization in practice of the dimension of Clean Environment was found to have been virtually unchanged over the first year of implementation.

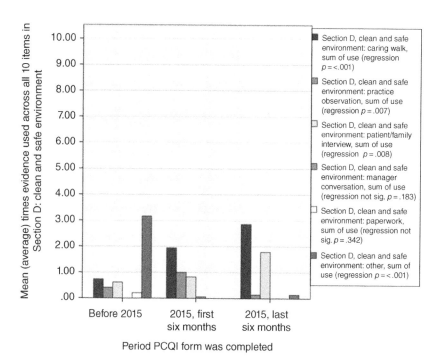

Figure 18.7 Types of evidence used for Section D, environment, 2012–2015.

Continuity of Care

Examination of evidence of Continuity of Care revealed over time that the caring walk and "other" proved the best way to observe the eight different actions of Continuity of Care outlined in Section E of the PCQI. In the last 6 months of 2015 cumulatively, Continuity of Care was observed over six times, which means Continuity of Care doubled from before 2015 to the end of 2015. As Figure 18.8 shows, statistically significant improvements were observed via the data collection methods of the caring walk ($p = <.001$) and other ($p = .001$).

Clinical Excellence

Examination of evidence of Clinical Excellence revealed that eventually the caring walk, practice observation, and "other" proved the best way to observe the nine different actions of Clinical Excellence outlined in Section F of the PCQI. As Figure 18.9 shows, statistically significant improvements were

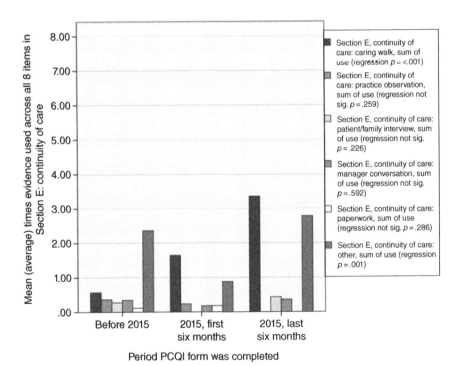

Figure 18.8 Types of evidence used for Section E, continuity of care, 2012–2015.

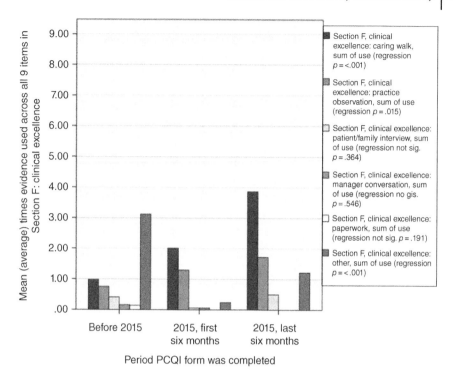

Figure 18.9 Types of evidence used for Section F, clinical excellence, 2012–2015.

observed primarily via the data collection method of the caring walk, where Clinical Excellence quadrupled in frequency over time ($p = .001$). Using the data collection method of practice observation, Clinical Excellence about doubled in frequency ($p = .015$). The use of the observation method of "other" decreased over time, again suggesting that Quality Champions learned what methods were best for collecting evidence of CBAS.

After the first year of implementation, though the PCQI showed positive results, it was determined that we would need to develop an instrument that could give us a clearer, more easily understood picture of the extent to which CBAS was permeating the culture. While all of the data scientists could see that the results of the PCQI were excellent, the results did not translate very well into layman's terms. We wanted more evidence to strengthen the argument that what staff members and leaders were reporting experientially was actually occurring.

These problems were solved by the development of the Operations of CBAS Assessment and use of the Healthcare Environment Survey (HES).

Effectiveness of CBAS Discovered by Using the Operations of CBAS Assessment and the HES

Four years into the implementation of CBAS, 346 nurses were asked to report whether they had or had not attended the 3-day CBAS training; if they had attended, they were asked to report how long it had been since they had attended. As Table 18.2 shows, 4 years into implementation of CBAS, fewer than 32% of the nurses had received the 3-day CBAS training.

Table 18.2 Nurse attendance of CBAS training and time since training.

How recently CBAS training was completed	Number of nurses	Percentage of nurses
Within the last 12 months	26	7.5
At least 1 year ago, but less than 2 years	29	8.4
At least 2 years ago, but less than 3 years	24	6.9
At least 3 years ago, but less than 4 years	15	4.3
At least 4 years ago, but less than 5 years	13	3.8
Five or more years	3	.9
Not applicable, I have not had CBAS training	236	68.2
Total	346	100.0

The peak score was for nurses who had received training between 3 and 4 years prior, suggesting that it may take time to learn the culture of CBAS.

The Operations of CBAS Assessment total score (all 17 items combined) was highest for nurses who had attended the CBAS course between 3 and 4 years or 4 and 5 years ago, which was about half of the 110 who had attended at all. It is interesting to note that the peak score was for nurses who had received training between 3 and 4 years prior, suggesting, among other things, that it may take time to learn the culture of CBAS. Operations of CBAS scores for all 346 nurses are noted in Figure 18.10. Notice that nurses who had not attended the 3-day CBAS training have relatively low scores, but nurses who attended the

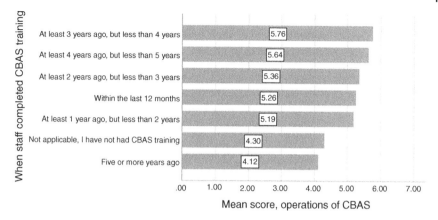

Figure 18.10 Operations of CBAS mean scores for nurses trained in CBAS.

3-day CBAS training 5 or more years previously scored even lower.

It is worth noting that the two items from the PCQI that related most strongly to quality and caring were those in which individuals agreed to the statements "I understand how the PCQI is used to improve patient care" and "I remember using the PCQI well in my CBAS training." These two items had a correlation with staff report of caring and quality of .316 ($p < .001$) and .324 ($p < .001$), suggesting that as knowledge of using the PCQI increased, so did staff perception of their own increased caring and quality.

How Knowledge of CBAS Affected Caring as Perceived by Staff

The 110 nurses who attended CBAS were then examined more deeply, and it was found that nurses having deeper understanding of the 17 concepts measured in the Operations of CBAS Assessment also reported higher caring and quality scores as measured by the Healing Compassion Survey—7 Cs NHS Scotland (Staff Version) (HCS-S), and the 56-item Healthcare Environment Survey (HES), which measures nurse job satisfaction. Thirteen items from the HCS-S had a statistically significant relationship with the Operations of CBAS Assessment, which means the more people understood CBAS, the more they perceived that they gave high quality care, and the more they liked their jobs. It was shown that when understanding of CBAS improved, so did the employee's perception of the following 13 caring and quality items:

For the patient you just took care of . . .
1) Were you supportive?
2) Were you caring?
3) Did you demonstrate your commitment to quality?
4) Were you an attentive listener?
5) Were you informative?
6) Did you display a "can do" attitude at every opportunity?
7) Were you understanding?
8) Did you take responsibility to do the job well?
9) Were you personal (treated the patient kindly and as an individual)?
10) Were you respectful of the patient?
11) Were you technically skilled?
12) Did you work effectively with others in teams?
13) Were you comforting?

Even with improved understanding of CBAS, however, the following five caring and quality items did not improve:

For the patient you just took care of . . .
1) Did you make sure you had everything he/she needed in his/her environment of care?
2) Were you able to encourage the patient to make decisions about their care/treatment?
3) Were you clinically competent?
4) Were you able to ensure that his/her faith or belief needs were met?
5) Has he/she consistently been part of your patient assignment?

More details on what specific aspects of CBAS correlated with caring as measured by the Healing Compassion Survey—7 Cs NHS Scotland (Staff Version) are noted in Appendix BB.

How Knowledge of CBAS Affected Job Satisfaction

When examining the 17 items from the Operations of CBAS Assessment, it was found that 16 of the 17 items had a statistically significant positive relationship with job satisfaction, meaning that as understanding of CBAS improved, so did job satisfaction. Only item 17, noted near the bottom of Table 18.3,

Table 18.3 Correlations of every item of the Operations of CBAS Assessment instrument with job satisfaction.

		HES total score
CBAS Operations total score	Pearson correlation	.625
	N	96
CBASq 1_The senior charge nurse is part of the support I feel is helpful to successfully implement the action plans we developed in the CBAS training	Pearson correlation	.664
	N	103
CBASq 2_The senior charge nurse I work with inspires me to apply what I learned during the CBAS training in patient care	Pearson correlation	.556
	N	104
CBASq 3_There are enough resources for me to feel we will be able to implement CBAS action plan on my ward	Pearson correlation	.528
	N	102
CBASq 4_Executive leaders are supportive of the change agenda that comes from our team action plan	Pearson correlation	.522
	N	103
CBASq 5_I feel there is adequate time for me to do what I feel I need to do to help implement our action plan following the CBAS training on my ward/department	Pearson correlation	.509
	N	102
CBASq 6_It was clear in our CBAS training how the PCQI could support our team to create action plans for change	Pearson correlation	.487
	N	100
CBASq 7_Executive leaders are aware of our team action plans	Pearson correlation	.484
	N	104
CBASq 8_I understand how the Person-centered Care Quality Instrument (PCQI) is used to improve patient car.	Pearson correlation	.483
	N	102
CBASq 9_I remember using the PCQI well in my CBAS training	Pearson correlation	.481
	N	102
CBASq 10_Executive board members continue to demonstrate support for the CBAS action plans following the CBAS implementation program	Pearson correlation	.476
	N	103
CBASq 11_The action plans following the CBAS implementation program have support from the executive board	Pearson correlation	.445
	N	104

(Continued)

Table 18.3 (Continued)

		HES total score
CBASq 12_The CBAS facilitators I work with are expert in helping me understand what behaviors I can change to provide the best patient care possible	Pearson correlation	.376
	N	101
CBASq 13_I am able to contact CBAS facilitators easily	Pearson correlation	.327
	N	104
CBASq 14_I believe CBAS is important for improving the care at this hospital	Pearson correlation	.318
	N	102
CBASq 15_The CBAS facilitators are always helpful	Pearson correlation	.308
	N	104
CBASq 16_I know who the CBAS facilitators are for my organization	Pearson correlation	.219
	N	104
CBASq 17_Coworkers I work with believe CBAS is important for improving the care at this hospital	Pearson correlation	.189
	N	102

Q: I have completed my CBAS training = Yes, completed CBAS and answered CBAS questions

did not have a statistically significant relationship with job satisfaction, using an alpha level of .05. The items of the Operations of CBAS Assessment are noted in Table 18.3 in rank order from those with the highest correlation to job satisfaction to those with the lowest.

Some of these correlations are particularly significant, indicating that as nurses understand CBAS more, they are more satisfied with their jobs. Of greatest interest, however, is the correlation between job satisfaction as a whole, as measured by the HES, and knowledge of CBAS, as measured by the Operations of CBAS Assessment, is .625 ($p = <.001$) which means that 62% of the time, when understanding of CBAS improves, so does job satisfaction.

The Operations of CBAS Assessment provided consistent evidence that as the behaviors named in the assessment improved, so did caring, quality, and job satisfaction, as reported by the staff. As noted earlier, 13 of the 18 items in the Healing Compassion Survey—7 Cs NHS Scotland (Staff Version) improved as operations of CBAS improved, and this finding was statistically significant.

The strongest argument for the impact of CBAS, however, is the statistically significant relationship found between the Operations of CBAS Assessment scores and the HES job satisfaction scores. Deeper examination using regression analysis would likely show that implementation of CBAS not only *relates* to job satisfaction, but actually *predicts* job satisfaction. The dramatic and consistent relationships between the behaviors named in the Operations of CBAS Assessment and the improved outcomes of caring, quality, and job satisfaction, which deepen as CBAS becomes more integrated into the culture, have served to propel this organization deeper into research to understand how interventions such as CBAS improve other outcomes for both patients and staff members.

Implementation of CBAS not only relates to job satisfaction, but actually predicts job satisfaction.

Action Planning

Managers and staff members who reviewed the PCQI, Operations of CBAS Assessment, and HES data every year gained confidence in the data produced from the instruments. Each ward reviewed their ward-specific data with Dr. Nelson to ensure the findings were properly understood and subsequently applied in action planning. No ward or year was the same and thus data was used for establishing, at the ward level, which dimensions of CBAS were consistently evident in practice and which dimensions needed to be addressed. Deep understanding of the data helped with staff engagement in action planning and prioritization of resources by management.

Operational refinements were made on all wards. What follows is an example of what operational changes were made on two postsurgical wards that had struggled with productivity and a perceived high level of workload. These wards were called 2 East and 2 West. Unit 2 East was open mostly Monday–Friday but was sometimes required to stay open on weekends for patients who could not be discharged, and 2 West was open 7 days a week. Both wards did elective surgeries. The focus of the action

Deep understanding of the data helped with staff engagement in action planning and prioritization of resources by management.

plans for both wards was workload, as staff from both wards reported frustration with heavy workload, and administration was concerned about improving productivity. The organization was not budgeted to hire additional staff, so an examination of the processes of work using just the existing staff and budget was undertaken by staff and leaders on the ward. The action plans the people on these wards developed helped decrease perception of high workload with concurrent improvement in productivity and other quality outcomes.

Ward 2 East (Post-Surgical Care): Action Plans and Outcomes

After staff members and leaders reviewed data in 2016, the following action plans were put into place, and outcomes were realized by the time data was collected, analyzed, and presented in 2019. This ward focused on workflow and how staff members were used. There were no staff members added during this time; instead, outcomes were achieved by redesigning how the work was performed using collaboration and rearrangement of the processes of work. The following actions were taken:

1) Ward 2 East noted the specialties of the nurses on their ward and on 2 West. Wards 2 East and 2 West then collaborated to ensure that each ward was getting the surgical patients who best matched the nurses' specialties.
2) Staff learned how to: (a) collect data on caring and quality using the PCQI, and (b) use data from the PCQI as evidence to compare what was really happening operationally to the organization's standard of care and quality.
3) All nurses were cross-trained to work on both 2 East and 2 West.
4) A higher degree of collaboration between 2 East and 2 West was established, to help with patient flow.
5) Ward 2 East was designated a fast-flow ward, meaning they accepted people needing procedures that were more likely to require a shorter stay. Sometimes they do need to take patients requiring longer stays, but the effort was to keep people requiring a shorter length of stay on 2 East.
6) Both wards kept "unfunded beds" (beds for which there are no staff members employed) closed.

7) Ward 2 East designated two beds for patients requiring higher levels of monitoring.

8) Both wards increased use of Advanced Nurse Practitioners to help expedite care.

9) Skill mix changed from 80% RNs and 20% non-registered nurses to 65% RNs and 35% non-registered nurses, and additional training was provided for non-registered nurses so they could do more of the procedure work RNs were doing, such as admitting patients, taking vital signs, and drawing blood.

10) Staff members met on several occasions to help solve problems of workload.

11) When 2 East opened each Monday, they started filling beds sequentially, so as the ward filled up, it filled up from one end of the ward to the other. Because of the shape of the ward, this meant fewer staff members were required, as they could see and reach more patients with less work. This is in contrast to alternating admissions to the two hallways in the ward that formed a half square, which spread patients out unnecessarily. Filling up in this way also helped form cohorts of patients which meant that all the allocated patients had the same recovery goals which facilitated consistency of approach and optimized nurses' time to care.

Ward 2 West (Postsurgical Care): Action Plans and Outcomes

After staff members and leaders reviewed data in 2016, the following action plans were put into place, and outcomes were realized by the time data was presented in 2019. This ward also worked on workload by adjusting workflow and how staff members were used:

1) Ward 2 West focused on team building. Staff members now rotate between 2 East and 2 West, which has helped the two staffs understand each other's ward area better. The rotation has also helped the senior charge nurse get to know staff members better.

2) A discharge lounge was created to increase comfort for patients and families and help get the patients in and out

more quickly. It also helped clinicians focus on the last elements of care just prior to discharge to ensure that the family understood the recovery goals.

3) Ward 2 West collaborated daily with 2 East to help organize patient placement, which helped overall flow.

4) A decision was made to keep patients with complex comorbidities on one ward (2 West) rather than on both wards. This allowed 2 East to close on weekends and decreased length of stay (LOS) from five days to one to two days across both wards for non-complex patients. This enabled nurses to concentrate on specific elements of the recovery pathway, which meant that patients started meeting their recovery goals much faster. The focus of the "fast-flow" nurses was on re-enablement, recovery, and early discharge. This new approach had a radical effect on LOS.

It was the observation of Dr. Nelson, in the 2019 presentation, that staff members and leaders were exuberant in their presentation of action plans and the resultant improvement in total job satisfaction. It was rewarding to see how staff members learned to use data for operational refinement that not only improved outcomes such as length of stay, but also improved the experience of work as evidenced by the improved scores in job satisfaction.

Discussion

The most important aspect of the research outlined in this chapter, as of 2019, was development and validation of instruments to capture caring as taught by CBAS and mandated by the NHS of Scotland. These instruments allow the effectiveness of CBAS implementations to be tested with measurement instruments specified for the context of CBAS, using questions and statements that were derived directly from the description and objectives of the CBAS program. Scientific validation and testing of the instruments caused the people involved in the study and action planning to have greater confidence in the data. Studies conducted using instruments that were sensitive enough to detect how the concepts of CBAS related to targeted outcomes revealed the following:

1) Patient's reported consistently high levels of caring and quality. (This was reviewed in detail in Chapter 17.)

2) Taking the CBAS training caused a majority of staff members to report they were more consistently (a) enacting caring and quality in patient care, and (b) experiencing greater job satisfaction. (It is likely that job satisfaction increased because staff members learned how to enact Care and Collaboration with ward colleagues after the PCQI identified Care and Compassion and Collaboration as dimensions requiring focused implementation and monitoring.)

3) Processes of work were improved at the ward level.

These findings, using instruments specified for the context of a CBAS implementation in Scotland, all point toward positive outcomes for patients, staff members, operations, and the financial health of the organization.

It is also worth noting that these improvements occurred with only 32% of the staff trained in CBAS. What would the impact be if 100% of the staff had been training in CBAS within the last 3–4 years and subsequently reported deep knowledge of the operations of CBAS as measured by the Operations of CBAS Assessment? It is exciting to ponder how an increase in CBAS training might positively affect the patients, staff members, and financial outcomes of this organization!

Epilogue: Imagining What Is Possible
John W. Nelson

Several years ago, while working in a 350-bed community hospital in the metropolitan area of Detroit, Michigan, I had an experience that showed me just how transformative data can be. I was in a meeting with two of the organization's housekeepers, both nearing retirement age, to review their department's results in job satisfaction; clarity of self, role, and system; and caring moments. They were from a high-performing department with consistently high satisfaction scores. These two people were highly engaged and laser focused on creating caring moments for others through their work. This organization believed that everyone can create a caring moment and thus had all staff members trained in Relationship-Based Care® (RBC) (Creative Health Care Management, 2017; Koloroutis, 2004) to ensure that every person in the organization was focused on the work of ensuring that patients and families are held in the center of a healing culture.

After reviewing their results, they were not surprised by their excellent scores, as they already knew they did good work. However, they shared that any time there was a dip in their scores (even if their scores were still, by any standard, exemplary), they were reprimanded by their department manager. As you can imagine, they found this distressing and unnecessary. I told these two housekeepers not to be discouraged because a dip is meaningless when using Gaussian mathematics, which is what had been used to measure the effectiveness of their work. I informed them that Gaussian mathematics is what

Using Predictive Analytics to Improve Healthcare Outcomes, First Edition.
Edited by John W. Nelson, Jayne Felgen, and Mary Ann Hozak.
© 2021 John Wiley & Sons, Inc. Published 2021 by John Wiley & Sons, Inc.

the science of statistics is premised on, and that it uses the mean score, which is the average of high and low scores. Thus, when following mean scores over time, there will always be a dip, no matter how good you are, because of the mathematics being used. They found this very interesting and wrote the term "Gaussian mathematics" in their notes.

As they were so engaged in the process and so appreciative of this new information, I decided to give them some more tools to facilitate a new kind of dialog with their manager the next time they were reprimanded for a dip in their scores. I told them to respond to their manager next time by stating, "The line graph you are showing me is Gaussian mathematics, and the overall trend of improvement for this department is steadily increasing. We have had only one drop in that trend, and thus, we have not yet established a new pattern. This single drop could be what is called, in Gaussian mathematics, regression to the mean. If this pattern continues, then we do agree that we should address this as an issue." The two housekeepers were delighted with this new information and wrote it down, word for word.

About a month later, I had a phone call from the chief nursing officer asking me if I had presented data to these two housekeepers because they were very clear in their understanding of statistics when the manager of housekeeping recently reprimanded them for a dip in their mean score. I laughed with delight to hear how data can empower when it is understood. We need more housekeepers—and truly everyone in healthcare—to understand and use data to improve their practice so they can create caring moments!

So what is next in your organization?

In recalling the analogy of the 2002 Oakland A's baseball team as described in the book and film *Moneyball* (Lewis, 2004), you may remember that the outcome variable they sought to change was not simply "number of games won." It was "on-base percentage." They understood that they could win more games if they simply increased the chances of getting a player to first base. We all want to be #1, but in healthcare, like in baseball, there are lots of fundamentals to get right in order to make it to the top.

Here are just a few things that are possible using predictive analytics:

1) Scheduling models that sort for the safest possible care team for your exact patient population on any given day.
2) Hiring models that help you select candidates with the relational and technical skills known to be most effective in a specific care environment.
3) Models to discover what caring behaviors reduce lengths of stay, prevent infections, and reduce patient falls.
4) Models to discover what levels of education or experience of clinicians are important to reduce hospital readmissions.
5) Models to discover whether providing alternative therapies to patients has a therapeutic/self-care effect on the clinicians who provide them.
6) Models to discover which caring behaviors of managers increase professional growth of staff, and decrease sick time and turnover.

The possibilities are endless.

Most of us agree that healthcare is getting more complex by the minute, but there are a lot fewer of us who believe that it is also getting more fun. I can tell you from experience that the best day you will ever have at work is the day you figure out how to save your organization a few million dollars while preventing life-threatening complications for 50 patients and their loved ones.

I can tell you from experience that the best day you will ever have at work is the day you figure out how to save your organization a few million dollars while preventing life-threatening complications for 50 patients and their loved ones.

Deepening use of predictive analytics and use of nonlinear methods of analytics will eventually make healthcare sing a beautiful song of improved efficiency, improved operations, and improved outcomes. However, when comparing the state of the predictive analytic "song" in healthcare to that of other industries, healthcare currently sounds more like a mumble. This work is possible for every organization, and, in fact, we are not far from a time when it will be considered unthinkable for a healthcare organization to forgo the ability to proactively manage risk.

Appendix A

Worksheets Showing the Progression from a Full List of Predictor Variables to a Measurement Model

This appendix provides a visual representation of the progression from a full list of predictor variables generated from the brainstorming of the people closest to the work to a model showing only the most significant variables, arranged into constructs.

Figure A.1 shows a full list of predictor variables generated by a staff council studying workload on a neurosurgical unit. Their story is recounted in the article "Measuring Workload of Nurses on a Neurosurgical Care Unit" (Nelson, Valentino, et al., 2015).

The work of arranging predictor variables into constructs is usually both messy and fun. Though shown in black and white, you can see in Figure A.2 how color-coding helped this group show which predictor variables had most in common.

The final figure in this appendix shows an alternate view of Figure 1.1 in Chapter 1, which shows the predictor variables arranged in a "spoke" formation around the variable of interest. In that figure, the proximity of the predictor variables to the center of the model shows how closely they relate to the variable of interest. In the model in Figure A.3, the predictor variables are numbered and you can see that they have been colored to show that they are in the same construct, but the rank of each variable' relationship to the variable of interest is not noted.

Since these are the variables actually identified as being able to be measured and what was finalized for measurement, this is

Using Predictive Analytics to Improve Healthcare Outcomes, First Edition.
Edited by John W. Nelson, Jayne Felgen, and Mary Ann Hozak.
© 2021 John Wiley & Sons, Inc. Published 2021 by John Wiley & Sons, Inc.

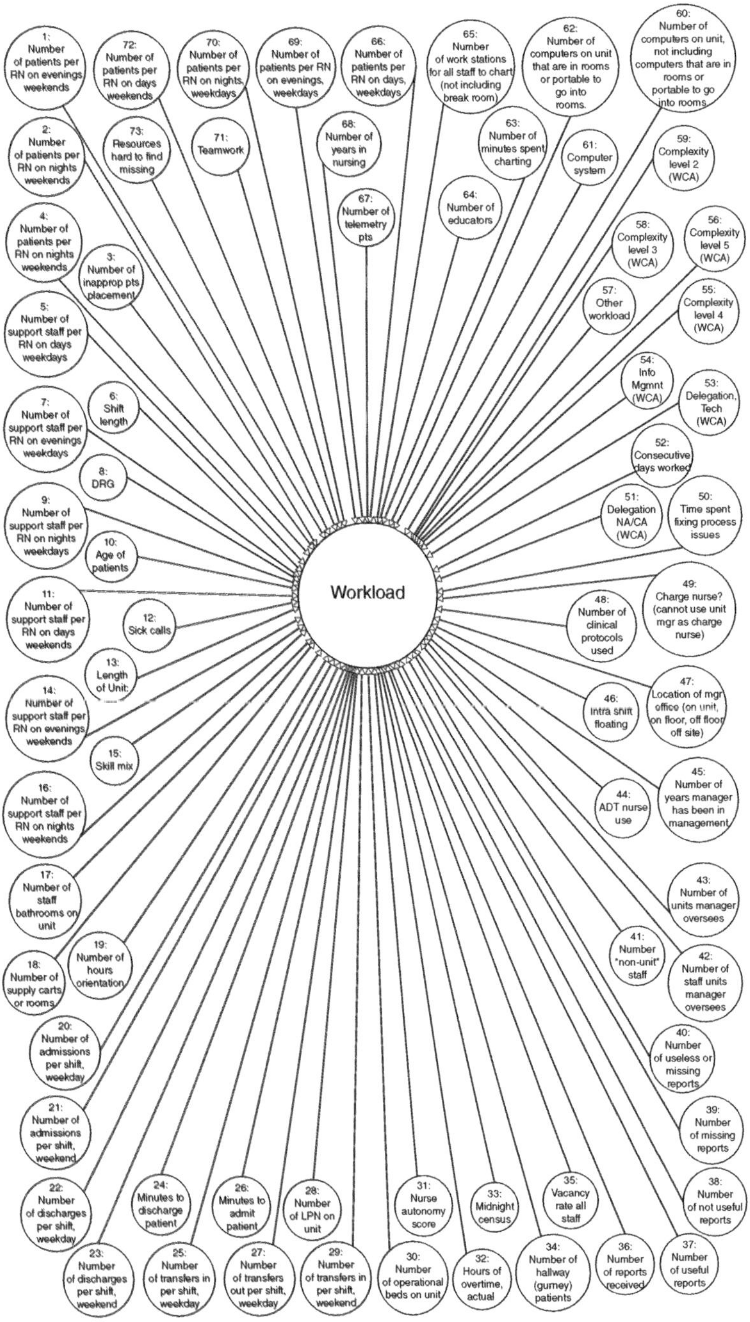

Figure A.1 Full list of predictor variables related to workload, the structural model.

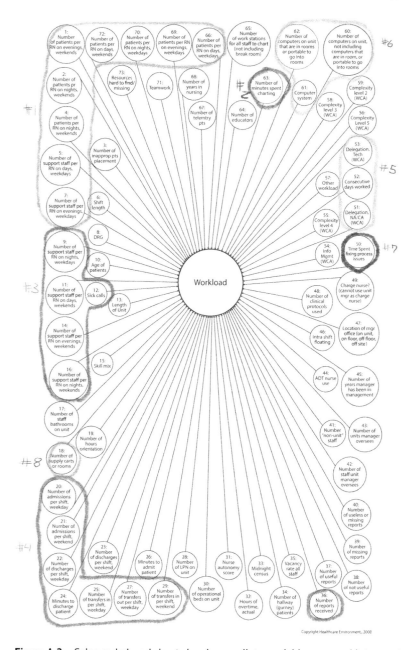

Figure A.2 Color-coded worksheet showing predictor variables grouped into constructs.

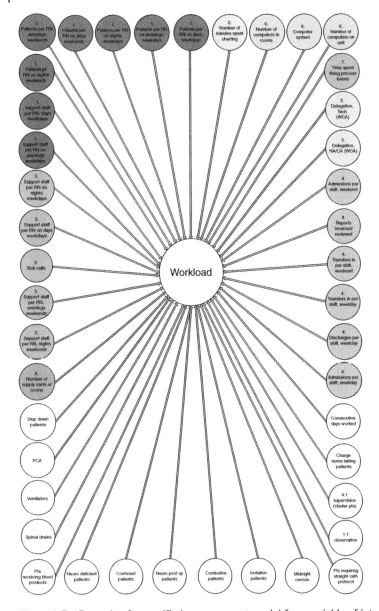

Figure A.3 Example of a specified measurement model for a variable of interest.

referred to as the measurement model. In this measurement model, the circles that are both colored and numbered were used to study the individual level of workload, while the uncolored and unnumbered circles were used to study the unit level of workload.

Appendix B

The Key to Making Your Relationship-Based Care® Implementation Sustainable Is "I_2E_2"

The critical component of any transformation process is a comprehensive and practical implementation plan that embeds the innovations and key concepts into daily work practices. The RBC model employs the formula for change outlined in the book I_2E_2: *Leading Lasting Change* (Felgen, 2007).

The I_2E_2 formula includes Inspiration (I_1), Infrastructure (I_2), Education (E_1), and Evidence (E_2). It provides a template for intentional planning specific to each desired innovation, shared vision, or goal. Here is a description of the elements of I_2E_2, beginning with its precursor, Vision:

Vision

The I_2E_2 model begins with a shared vision of what you'd like to create. No matter how large or how small the change, you'll need a clear vision of what you want to see happen in the organization or work area. Once your vision is determined and enhanced, refined, and affirmed by colleagues, and it becomes the shared vision for your group, four more elements—inspiration, infrastructure, education, and evidence—are considered:

Inspiration (I_1)

What inspires you and others to work toward the vision? Who else might be equally excited to be involved in helping this shared vision become a reality? Knowing you are not alone in this endeavor is important. What is it about this vision that will light people up, keep them energized, and make them want to participate? This is often the *why* behind the whole initiative.

Infrastructure (I₂)

For an implementation such as RBC to be launched and sustained, you must identify all the people who will be impacted as well as the structures and processes that need to be altered to reinforce the change and to make sure RBC is embedded in the day-to-day work. These structures and processes should be revised by staff councils (the people closest to the work) to help bring RBC to life at the tactical and operational levels.

Education (E₁)

Inevitably, any new initiative requires new knowledge and/or skills. People throughout the organization may need education on how to have successful meetings, how best to get the input of other team members, how to measure their success, and more. In later stages of the change, a new idea or practice may require staff, leader, or provider education before it is rolled out. Education needs may vary along the way, so your plan should anticipate these changes.

Evidence (E₂)

For any change to be successful, we must determine how we will know when our vision has become a reality. Early in the implementation process, councils must collect baseline data (which often exists already) of the current state. This data may be numbers (quantitative) or it may be narratives, opinions, and comments (qualitative). The collection of this baseline data puts you in the ideal position to reassess, after your new innovation is rolled out, to see whether it has had the intended impact. Seeing evidence of positive changes makes the entire effort rewarding and keeps staff council members inspired to make the next best innovation.

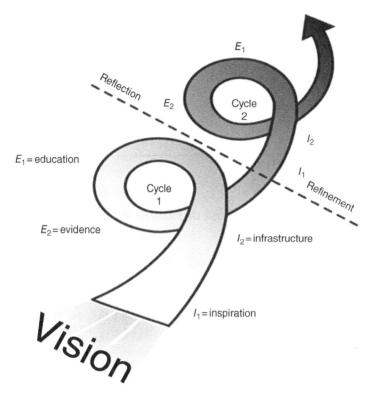

E_1

Reflection

E_2

Cycle 2

I_2

E_1 = education

I_1

Cycle 1

Refinement

E_2 = evidence

I_2 = infrastructure

I_1 = inspiration

Vision

Figure B.1   The I_2E_2 formula for leading lasting change.

As you can see in Figure B.1, central to the I_2E_2 model is reflection at regular intervals to appreciate progress, determine key success factors, and explore what more is needed to enhance the overall effort.

Designed to be revisited at critical junctures in the change process, this dynamic model becomes a way of being—not merely a checklist of tasks or a one-off exercise. It is practical and useful regardless of the scope of the expected change. It works for individuals planning specific changes in their practice, and it works for teams, departments, clinics, facilities, and systems. A comprehensive, easily applied description of I_2E_2 can be found in the book *I_2E_2: Leading Lasting Change* (Felgen, 2007).

Appendix C

Calculation for Cost of Falls

Cost calculations in this appendix are taken from a document prepared for use by a research team working on a falls reduction study at an organization we will refer to as Sample Community Hospital (SCH), 2015. The intervention was a predictive analytics study to identify what variables were predicting falls, in rank order. Below is the description of the cost calculation we used for falls.

In September 2015 we calculated the cost of falls using the following steps:

1) We used SCH's main website to find the number of admissions per year. Data from 2011 was reported on the SCH website, which was the most recent data available at the time of our calculation. We found there were 18,974 admissions at SCH in 2011.

2) We found that the average length of stay (LOS) for SCH on Hospital Compare was 4.3 days.

3) We multiplied 18,974 admissions by 4.3 days which came to 81,588.2 patient days.

4) Rate of falls, at the time of the study, was 4 per 1,000, which calculated to 326 falls for 81,588.2 patient days.

5) Cost for falls with serious injury is $13,316 (Wong et al., 2011). We used this number because costs vary greatly from state to state in the United States (Haddad, Bergen, & Florence, 2019), and these numbers were from a study analyzing the cost of falls in three Midwestern states. Total cost for all falls in 2011, if all falls were with serious injury, would be $4,341,016.

6) Cost for a fall with no injury is as low as $425 (Falen, Alexander, Curtis, & UnRuh, 2013). Total cost if none of the 326 falls were injurious would be $138,550.

7) Rate of injury (within the study site) at the time of the 2015 study was .36, so this is the number we used for our calculation. This number was slightly higher than other reports we have found with some indicating a rate of .25 of falls incurring injury.

8) Thus, when calculating .36 times 326, we can propose we had:
   a) A total of 117 falls with injuries in 2011, with a cost of $1,562,765.76.
   b) A total of 209 falls without injuries, with a cost of $88,825.00.
   c) Total combined (injury and no injury) for 326 falls was $1,651,590.76.

9) If falls decreased to zero for the year, we would save $1,651,59.76, based on data from 2011.

10) When comparing 2011–2015, the year we did the study (January–May), we had 170 falls on 7 units. That equals 34 falls per month, which would tally to 408 on the study units for the year. (Critical care did not participate. Only med/surg and rehab units participated.)
   a) 408 falls×.36 = 147, indicating that falls with injury during the study period incurred a cost of $1,957,452.00.
   b) 261 falls without injury during this time incurred a cost of 110,925.00.
   c) Total combined cost of falls on the units examined during the study incurred costs of $2,068,377.00.

Even though we were informed that we were down to "no falls" on study units, we knew falls were occurring in the critical care areas and thus estimated a cost savings of $1,600,000 for reduction of falls.

Appendix D

Possible Clinical, Administrative, and Psychosocial Predictors of Readmission for Heart Failure in Fewer Than 30 Days After Discharge

The final predictors found to be statistically significant in the prediction of readmission for heart failure in fewer than 30 days after discharge, in at least 1 of 25 models as reported in the literature review by Mahajan, Heidenreich, Abbott, Newton, and Ward (2018), include the following variables, organized into the domains of (a) clinical, (b) administrative, and (c) psychosocial. The list included 27 clinical variables, 37 administrative variables, and 11 psychosocial variables. All 25 models included clinical variables (Mahajan et al., 2018). Several models also used administrative data that is readily available within organizations, but these models were shown to have low predictive ability (Mahajan et al., 2018). The addition of psychosocial factors to predictive models is new, and opinion varies as to their importance in predicting readmission (Mahajan et al., 2018). More research is needed to understand whether psychosocial factors help predict readmission.

1) The list of clinical variables included 27 variables that varied in the operational definition (how it was measured). They are as follows:
  1.1   Respiratory rate (2 models)
  1.2   Heart rate (3 models)
      1.2.1   Heart rate
      1.2.2   Heart rate less than 80 beats per minute
  1.3   SBP (9 models)
      1.3.1   SBP less than 100

1.17.3 Loop diuretics
1.18 Admission nitrates (1 model)
1.19 Admission digoxin (1 model)
1.20 Discharge lipid lowering therapy (1 model)
1.21 Discharge angiotensin-converting enzyme/angiotensin receptor blocker (1 model)
1.22 Beta-lactam antibiotic on discharge (1 model)
1.23 Number of medications (2 models)
  1.23.1 Number of medications using ranges
  1.23.2 Number of medications upon discharge
1.24 ST-T changes in EKG (1 model)
1.25 Echocardiography (1 model)
1.26 Cardiac catheterization (1 model)
1.27 Pulmonary congestion per radiography (1 model)
2) The list of administrative variables included 37 variables that varied in the operational definition (how it was measured). They are as follows:
  2.1 Age (5 models)
    2.1.1 Age per 10 years
    2.1.2 Age discretized
    2.1.3 Age more than 55 years
  2.2 Sex (4 models)
    2.2.1 Sex
    2.2.2 Women (versus men)
  2.3 Race (3 models)
    2.3.1 Black race
    2.3.2 Black race (versus White)
    2.3.3 Black and Hispanic (versus White)
  2.4 Waist-hip ratio (1 model)
  2.5 Carlson Comorbidity Index (2 models)
  2.6 History heart failure (6 models)
    2.6.1 Prior heart failure
    2.6.2 Prior heart failure for more than 10 years
  2.7 Cardiomyopathy (3 models)
    2.7.1 Idiopathic cardiomyopathy
    2.7.2 Ischemic cardiomyopathy
  2.8 Coronary heart disease (2 models)
  2.9 Ischemic heart disease (2 models)
  2.10 Myocardial infarction (2 models)
  2.11 Valvular heart disease (2 models)
  2.12 Vascular/circulatory disease (1 model)

2.13 Peripheral vascular disease (1 model)

2.14 Arrhythmias (3 models)

2.15 Diabetes mellitus (5 models)

2.16 Renal disease (4 models)

2.17 Chronic lung disease (4 models)

2.18 Cerebrovascular accident or transient ischemic attack (3 models)

2.19 Cancer (3 models)

    2.19.1 Cancer

    2.19.2 Lymphoma

2.20 Liver disease (3 models)

2.21 Injury and poisoning (1 model)

2.22 Comorbidities (2 models)

2.23 Prior cardiac surgery (2 models)

2.24 Implantable cardioverter-defibrillator placement during hospitalization (1 model)

2.25 History of percutaneous coronary intervention (2 models)

2.26 Insurance (3 models)

    2.26.1 Medicare

    2.26.2 Medicaid

2.27 Prior admission (7 models)

    2.27.1 Prior admission within 1 year

    2.27.2 Prior admission within 30 days

    2.27.3 Index admission via ED 6 a.m. to 6 p.m.

2.28 Use of rural hospital (1 model)

2.29 Use of telemetry during admission (1 model)

2.30 Hospital service (1 model)

2.31 Mechanical ventilation during admission (1 model)

2.32 Length of stay (3 models)

    2.32.1 Length of say

    2.32.2 Length of stay more than 5 days

2.33 Discharge (4 models)

    2.33.1 Discharge to skilled nursing facility

    2.33.2 Discharge disposition

    2.33.3 Discharge between 11 a.m. and 7 p.m.

    2.33.4 Discharge to specific zip code

    2.33.5 Number of discharges

2.34 Home healthcare services after discharge (1 model)

2.35 ED visits (4 models)

    2.35.1 ED visits in prior 6 months

    2.35.2 ED visits in prior 3 months

2.35.3 ED visits in prior 30 days

2.35.4 ED visit frequency in prior 6 months

2.35.5 ED length of stay

2.36 Outpatient visits (2 models)

2.37 Prevalence of heart failure hospitalization for reference population (1 model)

3) The list of psychosocial variables included 11 variables that varied in the operational definition (how it was measured). They are as follows:

3.1 Single marital status or lives alone (3 models)

3.2 Sedentary lifestyle (1 model)

3.3 Retired (1 model)

3.4 Number of home address changes (1 model)

3.5 History of missed clinic visits (1 model)

3.6 Use of health system pharmacy (1 model)

3.7 Residence census tract in lowest socioeconomic quintile (1 model)

3.8 Drug/cocaine or alcohol abuse (2 models)

3.9 Depression, anxiety, or psychiatric disorder (3 models)

3.10 Protein calorie malnutrition (1 model)

3.11 Functional issue (3 models)

3.11.1 Swelling issues

3.11.2 Bathing

3.11.3 Shortness of breath

3.11.4 Short Form Health Survey

Appendix E

Process to Determine Variables for Lee, Jin, Piao, & Lee, 2016 Study

The 10 variables identified as predictors for falls in the Auto-FallRAS included five patient variables, 3 environmental factors, and 2 medical interventions (Lee et al., 2016). The 4 steps they used to drill down from 4,211 variables and interventions to 10 variables and interventions are as follows:

Step 1: The authors of this study calculated what they called an information value (IV), which is the degree to which a variable has a relationship—typically a predictive relationship—with falls. If a variable had an IV of at least .1, indicating it had a relationship with falls, it was then selected for further evaluation of the strength of its relationship with falls. The higher the IV score, the stronger the relationship the variable had with falls. There were 1,144 of 4,211 variables (27.2% of the variables) with an IV greater than .1, and all other variables were eliminated from the study. Of the remaining variables, those having an IV of .1–.3 ($n = 647$) were labeled as having moderate explanatory ability, and variables with an IV greater than .3 ($n = 497$) were labeled as having excellent explanatory ability (Lee et al., 2016).

Step 2: The 1,144 variables were evaluated further using individual regressions for each independent variable. In this step, 100 variables were removed due to not having a statistically significant relationship with falls, leaving 1,044 variables for additional analysis in Step 3.

Step 3: Various logistic regressions, which explained the odds of a fall occurring for each variable studied, were used to study the remaining variables. This analysis was conducted in conjunction with the literature to ensure that the variables retained had empirical support beyond this study. Twenty-six variables were selected for the final multivariate logistic regression.

Step 4: The final 26 variables were screened using the literature to determine whether they would be included in the final model. In this process, an additional 16 variables were eliminated. The remaining 10 variables included patient factors, environmental factors, and medical interventions.

Patient Factors Included in the Final Model

- Age, especially for 65 years old or older
- Maximum pulse rate, specifically for 100 beats-per-minute or more
- Registration as severely ill patient
- Activity level
- Hyponatremia (low sodium)

Environmental Factors Included in the Model

- Length of hospital stay
- Medical department, with admission to oncology relating to more falls, followed by internal medicine, then surgical
- Type of room, specifically rooms with three beds or more, regardless of type of unit

Medical Interventions Included in the Model

- Average number of daily tests, with increased tests relating to more falls
- Duration of time the patient is on nervous and circulation medication, with longer medication periods being associated with more falls

Appendix F

Summary of National and International Heart Failure Guidelines

National Institute for Health and Care Excellence (NICE) Guidelines

- Acute: Information on diagnosing and treating acute heart failure is very similar to the information presented in the NICE chronic heart failure management guidelines. Information is much more succinct however, lacking in detail on in-hospital care. Variables associated with in-hospital care primarily were measured through medications and procedures administered. The document did emphasize a multidisciplinary approach to care and the importance of clinical reasoning for improvement of care (National Institute for Health and Care Excellence, 2014).
- Chronic: Thorough in terms of providing detailed information pertaining to different types of treatment options (medications, procedure, rehabilitation, etc.), diagnostic tools, and importance of multidisciplinary care teams. Administration of pharmaceutical products for managing heart failure was extensively covered, and variables related to patient-centered care (communication, medication compliance, follow-up, etc.) were also identified. System-level factors were also outlined in order to easily implement the guidelines in an organizational setting (National Institute for Health and Care Excellence, 2018).

European Society of Cardiology (ESC), Acute and Chronic Heart Failure Guidelines

- Very thorough in terms of pharmacological treatment options (how, when, why) and extremely similar to the guidelines presented by the NICE chronic HF literature. They do provide a list of contraindicated treatments as well, of which two were recommended for use in the NICE guideline. The ESC when compared to NICE, provides significantly more information on the different types of imaging used to characterize heart failure with corresponding decision-making guidelines. Additionally, they provide a class (effect) and level (of evidence) for the recommendations made which is absent from the NICE guidelines. References are made to studies conducted that support and oppose the interventions stated throughout the literature (Ponikowski et al., 2016).

American College of Cardiology Foundation (ACCF)/ American Heart Association (AHA), 2013 and 2017 Guidelines for Management of Heart Failure

- The 2013 document is very similar to ESC in terms of thoroughness of pharmacological and imaging options for diagnosis and management of HF. Similarly, categorized recommendations by class (effect) and level (of evidence) which were color coded exactly as what was presented in the ESC guidelines. Treatments were assigned to stages of HF from A through D as well. Provided list of contraindicated treatments which were identical to what was provided by the ESC guidelines (Yancy et al., 2013)

- The 2017 document had more levels (of evidence) than the 2013 document and provided updates since the 2013 guideline. Special recommendations were made of HF comorbidities: anemia, hypertension (new), and sleep disordered breathing (Yancy et al., 2017).

Appendix G

Crosswalk Hospital Tool and Guidelines

This is a crosswalk of the guidelines found in the literature with the specifics of the organization's customized tool. Get with the Guidelines® (GWTG), which is the tool provided by the American Heart Association, was the foundation of this organization's tool. There were a total of 121variables in this tool, and 35 of them were also found in at least one of the guidelines provided by the National Institute for Health and Care Excellence, the European Society of Cardiology, and the American College of Cardiology Foundation. This meant that 86 of the variables were unique to the data collection tool used by the organization. We combined these 86 variables with 98 variables found in the literature, for a total of 184 variables. What follows in Table G.1 is a crosswalk of the 184 variables.

Table G.1 Crosswalk of hospital tools and guidelines.

Variable	Step to identify source of data (guidelines or hospital tool)			Guidelines					Predictive models
	Step 1: Number from guidelines	Step 2a: Number of variables in hospital data collection tool	Step 2b: Variables in hospital tool but not in guidelines	NICE (chronic HF) tool	NICE (acute HF) tool	ESC (acute and chronic HF) tool, 2016	ACCF/AHA tool, 2013	ACC/AHA/HFSA tool, 2017 (focused update of ACCF/AHA, 2013)	In Mahajan et al. (2018) HF predictive model review. Number below reveals number of models that used variable
Dataset		1	1						
Encounter number		1	1						
Patient ID		1	1						
Physician/Provider NPI		1	1						
Cardiologist		1	1						
Arrival date		1	1						
Admit date		1	1						
Arrival format		1	1						
Date of birth		1	1						
Age									6
Race		1	1						3
Gender		1	1						4
Marital status									3

Retired	1
Sedentary lifestyle	1
Number of home addresses	1
History of missed clinic visits	1
Use of health system pharmacy	1
Residence census tract in lowest socioeconomic quartile	1
Drug/cocaine or alcohol abuse	2
History of depression or anxiety or major psychiatric disorder	3
Protein calorie malnutrition	1
Functional: disability due to hemiplegia/ paraplegia/ paralysis	1

(*Continued*)

Table G.1 (Continued)

Variable	Step to identify source of data (guidelines or hospital tool)			Guidelines					Predictive models
	Step 1: Number from guidelines	Step 2a: Number of variables in hospital data collection tool	Step 2b: Variables in hospital tool but not in guidelines	NICE (chronic HF) tool	NICE (acute HF) tool	ESC (acute and chronic HF) tool, 2016	ACCF/AHA tool, 2013	ACC/AHA/HFSA tool, 2017 (focused update of ACCF/AHA, 2013)	In Mahajan et al. (2018) HF predictive model review. Number below reveals number of models that used variable
Functional: swelling issues in past 2 weeks									1
Functional: bathing yourself									1
Functional: shortness of breath issues in past 2 weeks									1
Functional: SF-12									1
Hispanic ethnicity		1	1						
Payment source		1	1						
Payment service		1	1						
Point of origin for admission/visit		1	1						
Medical history	1	1		1			1	1	5

If yes to CKD, factors considered	1		1	
History of cigarette smoking (past 12 months)	1	1		
Known history of HF prior to this admission	1	1		8
Cardiomyopathy				3
Known history of coronary heart disease				2
Known history of diabetes				5
Known history of renal disease				4
Known history of COPD/Asthma				4
Known history of CVA/TIA				3
Known history of cancer				2
Known history of lymphoma				1
Known history of liver disease				3

(Continued)

Table G.1 (Continued)

Variable	Step to identify source of data (guidelines or hospital tool)			Guidelines					Predictive models
	Step 1: Number from guidelines	Step 2a: Number of variables in hospital data collection tool	Step 2b: Variables in hospital tool but not in guidelines	NICE (chronic HF) tool	NICE (acute HF) tool	ESC (acute and chronic HF) tool, 2016	ACCF/AHA tool, 2013	ACC/AHA/HFSA tool, 2017 (focused update of ACCF/AHA, 2013)	In Mahajan et al. (2018) HF predictive model review. Number below reveals number of models that used variable
Injury or poisoning									1
Known history of ischemic heart disease									2
Known history of cardiac surgery									2
Cardiac catheterization performed									1
Known history of myocardial infarction									2
Known history of valvular disease									2
Known history of vascular/circulatory disease									1

Known history of peripheral artery disease		2
Arrhythmias		3
History of angiogram		2
Total number of diagnoses		2
VS: Heart rate	1	3
VS: SBP	1	9
VS: DBP	1	
Respiratory rate	1	2
Serum creatinine	1	6
Serum creatinine units	1	
Serum creatinine now drawn	1	
Blood urea nitrogen (BUN)		8
Glomerular filtration rate (GFR)		2
Sodium		2
Potassium		1
Troponin		2

(Continued)

Table G.1 (Continued)

| Variable | Step to identify source of data (guidelines or hospital tool) | | | Guidelines | | | | | Predictive models |
	Step 1: Number from guidelines	Step 2a: Number of variables in hospital data collection tool	Step 2b: Variables in hospital tool but not in guidelines	NICE (chronic HF) tool	NICE (acute HF) tool	ESC (acute and chronic HF) tool, 2016	ACCF/AHA tool, 2013	ACC/AHA/HFSA tool, 2017 (focused update of ACCF/AHA, 2013)	In Mahajan et al. (2018) HF predictive model review. Number below reveals number of models that used variable
Serum albumin									1
Hemoglobin									6
Hematocrit									1
Heart failure diagnosis		1		1					
Atrial fibrillation (at presentation or during hospitalization)		1	1						
Atrial flutter (at presentation or during hospitalization)		1	1						
EKG performed						1	1		1
EKG QRS duration (milliseconds)	1	1							

EKG not available	1	1				
EKG QRS morphology	1	1	1	1	1	
No ST changes in EKG						1
Supporting exams	1		1	1	1	
Supporting tests	1		1	1	1	1
Radiographic pulmonary congestion	1					1
Was peak flow or spirometry monitored	1		1	1		
If checkmark applied to NT-proBNP	1		1	1		3
If NT-proBNP, 400–2,000 ng/L, saw specialist and TTE within 6 weeks	1					
If NT-proBNP, >2,000 ng/L, saw specialist and TTE within 2 weeks	1					

(Continued)

Table G.1 (Continued)

Variable	Step to identify source of data (guidelines or hospital tool)			Guidelines					Predictive models
	Step 1: Number from guidelines	Step 2a: Number of variables in hospital data collection tool	Step 2b: Variables in hospital tool but not in guidelines	NICE (chronic HF) tool	NICE (acute HF) tool	ESC (acute and chronic HF) tool, 2016	ACCF/AHA tool, 2013	ACC/AHA/HFSA tool, 2017 (focused update of ACCF/AHA, 2013)	In Mahajan et al. (2018) HF predictive model review. Number below reveals number of models that used variable
Prior to TTE and specialist referral, were contributing factors to high NT-proBNP considered (e.g. age, LVH, etc.)	1			1		1			
If TTE performed	1			1					
Organization of care includes	1			1	1	1		1	
Medications given	1	1		1	1	1	1		
Factors of importance in ACEI administration	1			1		1		1	
Factors of importance in ARB administration	1			1		1	1	1	

Factors of importance in beta blocker administration	1	1	1	1
Factors of importance in aldosterone antagonist administration	1	1	1	1
Factors of importance in Ivabradine administration	1	1		
Factors of importance in Sacubitril valsartan administration	1	1		
Factors of importance in diuretics administration	1	1	1	3
Discharged with lipid lowering therapy	1			1
Supportive services provided	1	1	1	

(Continued)

Table G.1 (Continued)

Variable	Step to identify source of data (guidelines or hospital tool)			Guidelines					Predictive models
	Step 1: Number from guidelines	Step 2a: Number of variables in hospital data collection tool	Step 2b: Variables in hospital tool but not in guidelines	NICE (chronic HF) tool	NICE (acute HF) tool	ESC (acute and chronic HF) tool, 2016	ACCF/AHA tool, 2013	ACC/AHA/HFSA tool, 2017 (focused update of ACCF/AHA, 2013)	In Mahajan et al. (2018) HF predictive model review. Number below reveals number of models that used variable
EF – quantitative	1	1		1	1	1	1	1	
EF not available		1	1						
EF – qualitative									
Documented LVSD		1	1						
LVF assessment		1	1						
Was the patient ambulating at the end of hospital day 2		1	1						
Was DVT prophylaxis initiated by the end of hospital day 2		1	1						
Was DVT or PE pulmonary embolus documented		1	1						

Influenza vaccination	1	1	1	1
Pneumococcal vaccination	1	1	1	1
Discharge date	1	1		
Discharge time	1	1		
Patient's discharge disposition on the day of discharge (for patients discharged on or after 04/01/2011)	1	1		3
If home, special discharge circumstances	1	1		1
Discharged between 11 p.m. and 7 a.m.				1
Discharged to specific zip code				1
Admission between 6 a.m. and 6 p.m.				1
Use of rural hospital				1
Use of telemetry				1

(Continued)

Table G.1 (Continued)

Variable	Step to identify source of data (guidelines or hospital tool)			Guidelines					Predictive models
	Step 1: Number from guidelines	Step 2a: Number of variables in hospital data collection tool	Step 2b: Variables in hospital tool but not in guidelines	NICE (chronic HF) tool	NICE (acute HF) tool	ESC (acute and chronic HF) tool, 2016	ACCF/AHA tool, 2013	ACC/AHA/HFSA tool, 2017 (focused update of ACCF/AHA, 2013)	In Mahajan et al. (2018) HF predictive model review. Number below reveals number of models that used variable
Use of hospital service									1
Mechanical ventilation during admission									1
Length of stay									3
Prior ED visits									5
Outpatient visits									2
If other healthcare facility		1	1						
Comfort measure only		1	1						
Discharge systolic blood pressure supine		1	1						
Discharge diastolic blood pressure supine		1	1						

	C1	C2	C3	C4	C5	C6
Plan of care was established prior to discharge	1		1			
Family was included in discharge planning	1		1			
All services were set up prior to discharge	1		1			
Plan of care	1		1	1		
Plan of care given to care team	1		1	1		
ACEI prescribed	1	1	1		1	
ACEI contraindicated	1	1				
Contraindications or other documented reason(s) for not providing ACEI	1	1	1			
ARB prescribed	1	1			1	1
ARB contraindicated	1	1				
Contraindications or other documented reason(s) for not providing ARB	1	1			1	

(Continued)

Table G.1 (Continued)

Variable	Step to identify source of data (guidelines or hospital tool)			Guidelines					Predictive models
	Step 1: Number from guidelines	Step 2a: Number of variables in hospital data collection tool	Step 2b: Variables in hospital tool but not in guidelines	NICE (chronic HF) tool	NICE (acute HF) tool	ESC (acute and chronic HF) tool, 2016	ACCF/AHA tool, 2013	ACC/AHA/HFSA tool, 2017 (focused update of ACCF/AHA, 2013)	In Mahajan et al. (2018) HF predictive model review. Number below reveals number of models that used variable
Beta-lactam antibiotic on discharge									1
Admission nitrates									1
ARNI prescribed	1	1							
ARNI contraindicated		1	1						
Reasons for not switching to ARNI at discharge		1	1					1	
Reason for switch to ARNI at discharge		1	1						
New onset heart failure		1	1						
NYHA class I		1	1						
NYHA Class IV		1	1						

	1	2	3	4	5
If yes (referring to row 64)	1	1			1
Not previously tolerating ACE I or ARB	1	1			
Anticoagulation therapy prescribed	1	1	1	1	1
If yes (referring to row 69), class	1	1		1	
Anticoagulation therapy contraindicated	1	1			
Beta blocker prescribed	1	1		1	
If yes (referring to row 95), class of beta blocker	1	1	1		
If yes (referring to row 95)	1	1			
Beta blocker contraindicated	1	1			
Contraindications or other documented reason(s) for not providing beta blockers	1	1			

(Continued)

Table G.1 (Continued)

| Variable | Step to identify source of data (guidelines or hospital tool) | | | Guidelines | | | | | Predictive models |
	Step 1: Number from guidelines	Step 2a: Number of variables in hospital data collection tool	Step 2b: Variables in hospital tool but not in guidelines	NICE (chronic HF) tool	NICE (acute HF) tool	ESC (acute and chronic HF) tool, 2016	ACCF/AHA tool, 2013	ACC/AHA/HFSA tool, 2017 (focused update of ACCF/AHA, 2013)	In Mahajan et al. (2018) HF predictive model review. Number below reveals number of models that used variable
Aldosterone antagonist prescribed	1	1					1	1	
Aldosterone antagonist contraindicated		1	1						
Contraindications or other documented reason(s) for not providing aldosterone antagonist at discharge	1	1				1			
Ivabradine prescribed	1			1		1			
If yes (referring to row 80), reason	1			1		1		1	

Ivabradine contraindicated	1		1			
Hydralazine nitrate prescribed	1		1	1	1	
Hydralazine nitrate contraindicated		1	1			
Hydralazine nitrate contraindicated or other reason for not giving		1				
Digoxin prescribed	1		1	1		1
Digoxin contraindicated	1		1			
Diuretics prescribed	1		1	1	1	
Diuretics contraindicated	1		1			
Amiodarone prescribed	1		1			
If yes (referring to row 91), consider	1		1			
Amiodarone contraindicated	1		1			

(Continued)

Table G.1 (Continued)

	Step to identify source of data (guidelines or hospital tool)			Guidelines					Predictive models
Variable	Step 1: Number from guidelines	Step 2a: Number of variables in hospital data collection tool	Step 2b: Variables in hospital tool but not in guidelines	NICE (chronic HF) tool	NICE (acute HF) tool	ESC (acute and chronic HF) tool, 2016	ACCF/AHA tool, 2013	ACC/AHA/HFSA tool, 2017 (focused update of ACCF/AHA, 2013)	In Mahajan et al. (2018) HF predictive model review. Number below reveals number of models that used variable
Omega-3 fatty acids prescribed	1						1		
Omega-3 fatty acids contraindicated	1						1		
Medication compliance	1		1						
Number of discharge medications									1
Number of medications									1
ICD placement									1
ICD therapy counseling	1	1	1				1		
ICD therapy reason for not counseling		1	1						

Documented medical reason(s) for not counseling	1			1		
ICD placed or prescribed	1		1		1	1
Reason for not placing or prescribing ICD	1	1				
Documented medical reason(s) for not placing or prescribing ICD therapy	1			1	1	1
CRT-D placed or prescribed	1		1		1	1
CRT-P placed or prescribed	1		1		1	1
Reason for not placing or prescribing CRT	1	1				
Documented medical reason(s) for not placing or prescribing CRT therapy	1			1		

(Continued)

Table G.1 (Continued)

| Variable | Step to identify source of data (guidelines or hospital tool) | | | Guidelines | | | | | Predictive models |
	Step 1: Number from guidelines	Step 2a: Number of variables in hospital data collection tool	Step 2b: Variables in hospital tool but not in guidelines	NICE (chronic HF) tool	NICE (acute HF) tool	ESC (acute and chronic HF) tool, 2016	ACCF/AHA tool, 2013	ACC/AHA/HFSA tool, 2017 (focused update of ACCF/AHA, 2013)	In Mahajan et al. (2018) HF predictive model review. Number below reveals number of models that used variable
Additional therapies performed or planned as a result of this admission	1		1	1		1	1	1	
Smoking cessation counseling given	1	1				1			
Activity level	1	1				1	1		
If HF patient is a women with childbearing potential	1	1	1	1					
Follow-up	1	1					1		
Symptoms worsening	1	1	1						
Diet (salt restricted)	1	1		1			1		

Medications		1				1
Weight monitoring		1	1			1
Waist–hip ratio						1
Peripheral edema						1
Carlson comorbidity index						2
Elevated jugular venous pressure						1
Follow-up visit scheduled	1	1		1	1	
Ensure changes to clinical record are understood by patient	1	1		1		
Ensure changes to clinical record are shared with multidisciplinary team	1	1		1		
Date of first follow-up visit		1	1			
Time of first follow-up visit		1	1			

(Continued)

Table G.1 (Continued)

Variable	Step to identify source of data (guidelines or hospital tool)			Guidelines					Predictive models
	Step 1: Number from guidelines	Step 2a: Number of variables in hospital data collection tool	Step 2b: Variables in hospital tool but not in guidelines	NICE (chronic HF) tool	NICE (acute HF) tool	ESC (acute and chronic HF) tool, 2016	ACCF/AHA tool, 2013	ACC/AHA/HFSA tool, 2017 (focused update of ACCF/AHA, 2013)	In Mahajan et al. (2018) HF predictive model review. Number below reveals number of models that used variable
Location of first follow-up visit		1	1						
Medical or patient reason for no follow-up appointment being scheduled		1	1						
Follow-up phone call scheduled	1	1					1		
Date of first phone call		1	1						
Referral to outpatient HF management program	1	1			1				
Referral to AHA heart failure interactive workbook		1	1						

Provision of at least 60 minutes of heart failure education by a qualified educator	1	1		1	1	1
Advanced care plan/surrogate decision maker documented or discussed	1	1				
Advance directive executed	1	1				
Give information to people newly diagnosed with HF	1	1	1			
Discharge medications	1	1				
Follow-up treatments and services are needed	1	1				
Procedures performed during hospitalization	1	1				
Reason for hospitalization	1	1				

(Continued)

Table G.1 (Continued)

Variable	Step to identify source of data (guidelines or hospital tool)			Guidelines					Predictive models
	Step 1: Number from guidelines	Step 2a: Number of variables in hospital data collection tool	Step 2b: Variables in hospital tool but not in guidelines	NICE (chronic HF) tool	NICE (acute HF) tool	ESC (acute and chronic HF) tool, 2016	ACCF/AHA tool, 2013	ACC/AHA/HFSA tool, 2017 (focused update of ACCF/AHA, 2013)	In Mahajan et al. (2018) HF predictive model review. Number below reveals number of models that used variable
Treatments and services provided	1	1							
Care transition record transmitted	1	1							
Additional comments	1	1							
Is home located in a region with high levels of smog	1			1					
Is home located where safe drinking water is accessible	1			1					
Care transition record includes	1						1		
Care transition record transmitted	1			1					

Primary team (primary nurse, primary physician, and team) should take over routine management of HF	1	1
ICD-9 principal diagnosis code	1	1
ICD-10-CM principal diagnosis code	1	1
Was this case sampled	1	1
During this hospital stay, was the patient enrolled in a clinical trial in which patients with the same condition as the measure set were being studied (i.e. AMI, CAC, HF, PN, PR, SCIP)	1	1

(Continued)

Table G.1 (Continued)

Variable	Step to identify source of data (guidelines or hospital tool)			Guidelines					Predictive models
	Step 1: Number from guidelines	Step 2a: Number of variables in hospital data collection tool	Step 2b: Variables in hospital tool but not in guidelines	NICE (chronic HF) tool	NICE (acute HF) tool	ESC (acute and chronic HF) tool, 2016	ACCF/AHA tool, 2013	ACC/AHA/HFSA tool, 2017 (focused update of ACCF/AHA, 2013)	In Mahajan et al. (2018) HF predictive model review. Number below reveals number of models that used variable
PMT used concurrently or retrospectively or combination	1	1	1						
Standardized order sets used		1	1						
Patient adherence contract/compact used		1	1						
Discharge checklist used		1	1						
Clinical reasoning	1			1	1				
Physician trained in heart failure	1			1					
Nurse trained in heart failure	1			1					
Pharmacist trained in heart failure	1			1					

Staff has received formal training in heart failure	1		
Interdisciplinary collaboration	1	1	1
Lead full review of patient's heart failure care	1	1	
Patient experience	1		
Communication must be kind and candid	1		
Reassess, record, and treat according to change in heart failure status	1		
Reassess, record, and treat after recent interventional procedure (e.g. ICD placement)	1		
Reassess, record, and treat when not responding to current treatment	1		
Palliative care	1	1	1

(Continued)

Table G.1 (Continued)

Variable	Step to identify source of data (guidelines or hospital tool)			Guidelines					Predictive models
	Step 1: Number from guidelines	Step 2a: Number of variables in hospital data collection tool	Step 2b: Variables in hospital tool but not in guidelines	NICE (chronic HF) tool	NICE (acute HF) tool	ESC (acute and chronic HF) tool, 2016	ACCF/AHA tool, 2013	ACC/AHA/HFSA tool, 2017 (focused update of ACCF/AHA, 2013)	In Mahajan et al. (2018) HF predictive model review. Number below reveals number of models that used variable
Frequency of readmissions									6
Readmit 30 days or less									1
Prevalence of heart failure hospitalization for reference population									1
NA (for annual HF program audit)	1			1					
Frequency	98	121	86	78	8	43	41	20	

Appendix H

Comprehensive Model of 184 Variables Found
in Guidelines and Hospital Tool

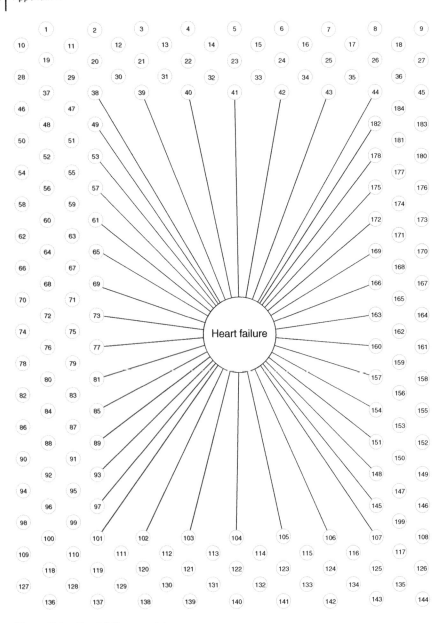

Figure H.1 Heart failure model.

Appendix I

Summary of Variables That Proved Insignificant After Analysis

The following variables showed no significant influence on whether heart failure patients were readmitted in fewer than 30 days after discharge.

Age

Age ranged from 22 to 96 years among the 214 patients for whom a full assessment was taken with a mean age of 71 and standard deviation of 13 years. Application of Pearson's correlation between age and number of readmission revealed no relationship between these two variables ($r = -0.040$, $p = 0.563$).

Gender

Among the 214 people in this study, there were 114 who people reported male and 100 people who reported female as their gender. Regression analysis of frequencies of admissions for these two groups revealed that gender did not play a role in readmission in fewer than 30 days (R^2 (change) $= .000$, $p = .863$).

Payment Type

There were 211 of 214 people who reported a primary payment type and 32 people who reported a secondary payment type. This analysis examined only primary insurance. Medicaid

($n = 73$) was used as the reference in the regression analysis. The Medicare–private/HMO ($n = 31$) and Private/HMO/other ($n = 8$) were combined into one group for analysis. Medicare ($n = 54$) was the final grouping used in the analysis. There were 45 people who reported no insurance or had no insurance source documented. Due to the "not documented" possibly including one of the other insurances, this group was not included in the analysis. All payments types and frequencies included in the analysis are noted in Table I.1.

Regression analysis revealed no relationship between frequency of admissions in fewer than 30 days and payment type (R^2 (change) $= .026, p = .256$).

Table I.1 Primary payment type frequency.

Primary payment	Frequency	Percent
Medicare	54	25.2
Medicaid	73	34.1
Medicare–private/HMO	31	14.5
Private/HMO/other	8	3.7
No insurance/not documented	45	21.0
Subtotal	211	98.6
Missing	3	1.4
Total	214	100.0

History of Smoking

There were 36 patients who reported a history of smoking, 169 who did not have a history of smoking, and 9 who did not respond to the question. History of smoking did not have a relationship to readmission in fewer than 30 days ($r = .040, p = .567$).

Vital Signs (Heart Rate and Blood Pressure)

Data for vital signs was available in datasets two and three. There were 118 patients for whom this data was available. Results revealed heart rate ranges from 53 to 146 with a mean rate of 83 (standard deviation 18). Heart rate did not have a statistically significant relationship with readmission in fewer than 30 days ($r = .052, p = .549$).

Systolic blood pressure had a negative relationship with readmissions in fewer than 30 days ($r = -.129$), but this relationship was not statistically significant ($p = .163$). Diastolic blood pressure had a slightly weaker negative relationship with readmission in fewer than 30 days ($r = -.122$) and was also not statistically significant ($p = .188$).

Serum Creatinine

There were 117 patients in datasets two and three for whom data related to serum creatinine was available. Serum creatinine values ranged from .5 to 10.61 with an average value of 1.99 (standard deviation of 1.86). There was no relationship found between serum creatinine and readmission in fewer than 30 days ($r = .007, p = .942$).

EKG

EKG QRS length in milliseconds (ms) and morphology of QRS were reported. There were 212 of 214 patients for whom data on QRS length was available. The range for QRS length in ms was 50–190 with a mean of 110.25 (standard deviation 29.24). Correlation of the QRS length with readmission in fewer than 30 days found no statistically significant relationship ($r = .049, p = .475$).

Examination of QRS morphology categories revealed that most patients in this study had a normal QRS ($n = 118$). All morphologies reported are noted in Table I.2. An ANOVA procedure was used to

Table I.2 QRS morphology groupings.

QRS morphology	Frequency	Percent
Normal	118	55.1
LBBB	18	8.4
RBBB	24	11.2
NS-IVCD	28	13.1
Paced	22	10.3
Total	210	98.1
System	4	1.9
Total	214	100.0

examine whether there was a difference in QRS morphology groupings as noted in Table I.2 for frequency of readmissions in fewer than 30 days. Results revealed there was no difference ($p = .832$) between any grouping. For example, those who had paced heart rhythm from a pacemaker had no more readmissions in fewer than 30 days when compared to those who had a right bundle branch block (RBBB). Frequency of each grouping is noted in Table I.2.

Follow-Up Visit Scheduled

There were 64 (30%) of the 214 patients who had a follow-up visit scheduled, 134 (63%) who did not, and 16 who did not report whether a visit was scheduled. Examination of a follow-up visit schedule in a regression of readmission in fewer than 30 days revealed that the presence or absence of the scheduling of a follow-up visit had no relationship with readmission (R^2 (change) $= .001, p = .615$).

ICD Codes 9 and 10

Responses for ICD Code 9 were available for only the early part of the study which meant that we had ICD Code 9 data for only 118 of the 214 people in the study. There were 3 main groups that were studied due to sample sizes that were large enough, including 42823 ($n = 39$), 42833 ($n = 33$), and 42843 ($n = 20$). All "other" codes in ICD 9 were used as reference. A regression equation revealed that ICD 9 had no relationship with readmissions in fewer than 30 day ($R^2 = .013, p = .679$).

ICD Code 10 information was provided in all three datasets. There were three codes large enough to group and compare, including I5023 ($n = 64$) which is acute on chronic diastolic heart failure, I5033 ($n = 49$) which is chronic diastolic heart failure, and I5043 ($n = 48$) which is acute on chronic combined systolic and diastolic heart failure. The 214 people in our study on whom full data was collected, reported 13 different ICD codes. Regression analysis revealed ICD10 code predicted 3.3% of the variance and was almost statistically significant ($p = .086$). When looking at the coefficients, ICD10 code I5033 had a statistically significant t-value (-2.180) which means that people with this ICD10 code were less likely to be readmitted in fewer than 30 days, and patients with ICD 10 codes I5023 and I5043 had higher early readmission rates. Both I5023 and I5043 include systolic heart failure, which is consistent with lower ejection fraction and LVSD, which were both statistically significant predicators of readmissions in fewer than 30 days in this study.

Hydralazine Nitrate Prescribed

There were 53 patients who reported being prescribed hydralazine nitrate and 154 who were not prescribed this medication. There were seven who did not report whether this medication was prescribed. Regression analysis revealed that prescribing of this medication had no relationship with readmission ($R^2 = .000$, $p = .991$).

Aldosterone Antagonist at Discharge

There were 48 patients who reported that an aldosterone antagonist had been prescribed for them at discharge. There were eight who did not respond to this item. Regression analysis of this item revealed having been prescribed an aldosterone antagonist had no relationship to frequency of readmission in fewer than 30 days ($R^2 = .002$, $p = .535$).

Beta Blocker Prescribed at Discharge

There were 182 (85%) patients who reported that they were prescribed a beta blocker at discharge and 25 who reported that they were not. There were seven who did not respond to this item. Regression analysis of this item revealed being prescribed a beta blocker at discharge had no relationship to frequency of readmission in fewer than 30 days ($R^2 = .000$, $p = .757$).

Anticoagulation Medication Prescribed at Discharge

There were 98 (46%) patients who reported that they were prescribed anticoagulation medication at discharge and 108 who reported they were not. There were eight who did not respond to this item. Regression analysis of this item revealed anticoagulation medication had no relationship to frequency of readmission in fewer than 30 days ($R^2 = .000, p = .959$).

ARNI Prescribed at Discharge

There were 62 (29%) patients who reported being prescribed ARNI medication at discharge and 140 who reported they were not. There were 12 patients who did not respond to this item. Regression analysis of this item revealed ARNI medication had no relationship to frequency of readmission in fewer than 30 days ($R^2 = .002, p = .419$).

ARB Prescribed at Discharge

There were 29 (14%) patients who reported that they had been prescribed ARB at discharge, and 178 who reported they had not. There were seven who did not respond to this item. Regression analysis of this item revealed that ARB medication had no relationship to frequency of readmission in fewer than 30 days ($R^2 = .007, p = .231$).

ACE Inhibitor Prescribed at Discharge

There were 67 (31%) patients who reported that they were prescribed an ACE inhibitor at discharge and 140 who were not. There were seven who did not respond to this item. Regression analysis of this item revealed that prescription of an ACE inhibitor had no relationship to frequency of readmission in fewer than 30 days ($R^2 = .007, p = .226$).

Pneumococcal Vaccination

Eighty-nine of the patients received a pneumococcal vaccination either in the hospital ($n = 17$; 8%) or prior to hospitalization ($n = 72$; 34%). Fifty-six patients (26%) did not receive the vaccine. All responses to this item are noted in Table I.3.

Table I.3 Did the patient receive pneumococcal vaccination?

Did patient receive pneumococcal vaccine?	Frequency	Percent
Pneumococcal vaccine was given during this hospitalization	17	7.9
Pneumococcal vaccine was received in the past, not during this hospitalization	72	33.6
Documentation of patient's refusal of pneumococcal vaccine	56	26.2
None of the above/not documented/UTD	66	30.8
Subtotal	211	98.6
Missing	3	1.4
Total	214	100.0

These two groups were compared using a regression equation to see whether receipt of the vaccination decreased readmission rates in fewer than 30 days post discharge. Results revealed that receipt of the vaccine did not have a statistically significant relationship with readmission in fewer than 30 days (R^2 .007, $p = .303$). Examination of the coefficients revealed that those who did receive the vaccine were readmitted more frequently in fewer than 30 days when compared to those who did not receive the vaccine ($t = 1.035$; $p = .303$), which was not statistically significant.

Influenza Vaccination

There were 85 of the patients who received an influenza vaccination either while in the hospital ($n = 17$; 8%) or prior to hospitalization ($n = 68$; 32%). Fifty-seven patients (27%) refused the vaccine, and three (1%) had contraindications. (See Table I.4.)

Those who received a vaccine and those who refused were compared using a regression equation to see whether receipt of the vaccination related to readmission in fewer than 30 days post discharge. Results revealed that receipt of the vaccine did not have a statistically significant relationship with readmission in fewer than 30 days ($R^2 = .010$, $p = .248$). Similar to those who received the pneumococcal vaccine, examination of the

Table I.4   Did the patient receive influenza vaccine?

Did patient receive influenza vaccine?	Frequency	Percent
Influenza vaccine was given during this hospitalization during the current flu season	17	7.9
Influenza vaccine was received prior to admission during the current flu season, not during this hospitalization	68	31.8
Documentation of patient's refusal of influenza vaccine	57	26.6
Allergy/sensitivity to influenza vaccine or if medically contraindicated	3	1.4
None of the above/not documented/UTD	66	30.8
Total	211	98.6
Missing	3	1.4
Total	214	100.0

coefficients revealed that those who received the influenza vaccine were readmitted more frequently in fewer than 30 days when compared to those who did not receive the vaccine ($t = 1.161; p = .248$), which was not statistically significant.

Appendix J

Summary of Inconclusive Findings

For a variety of reasons, the following variables produced inconclusive findings: the patient's race, ethnicity, smoking cessation, and activity level.

Race

There were two groups with adequate sample sizes for comparison: white ($n = 99$) and black or African American ($n = 74$). Patients whose race was listed as "unable to determine" were not included for comparison. American Indian or Alaska Native ($n = 2$) and Asian ($n = 4$) had sample sizes too small to include for comparison. (See Table J.1.)

Table J.1 Frequency of race.

Race	Frequency	Percent
Black or African American	74	34.6
American Indian or Alaska Native	2	.9
Asian	4	1.9
White	99	46.3
Unable to determine (UTD)	35	16.4
Total	214	100.0

Regression analysis revealed no relationship between race (white coded as 0 and black coded as 1) and frequency of admissions in fewer than 30 days ($p = .765$) or fewer than 90 days (R^2 (change) $= .001$, $p = .805$).

Ethnicity

Due to 161 of the 214 being labeled "No/unable to determine," the risk of error was too great, and this variable was not examined.

Smoking Cessation

There were 196 people who reported "yes" to whether smoking cessation was discussed, and 10 who reported "no." Due to the small sample size of people reporting "no" and almost all reporting "yes," this variable was not examined.

Activity Level

Similar to addressing smoking cessation, almost all responses to whether activity level was discussed were "yes" ($n = 201$); five people reported "no." This variable was not examined.

Appendix K

Nine Tools for Measuring the Provision of Quality Patient Care and Related Variables

The researchers in the study outlined in Chapter 10 of this book accessed six tools to assess caring—three to assess caring as perceived by patients and three to assess caring as perceived by staff members. This appendix also describes three more tools which were used by these researchers to measure self-care, nurse job satisfaction, and clarity of self, role, and system.

Three Tools for Assessing "Caring for the Patient" As Perceived By the Patient

There were three tools available to measure patients' perceptions of care.

Tool 1: The Caring Factor Survey (CFS)

The Caring Factor Survey (CFS) was used to measure the patients' perception of the 10 caring behaviors in Watson's Theory of Transpersonal Caring (2008a), using a 7-point Likert scale, ranging from strongly disagree (1) to strongly agree (7). Higher mean scores of the 10 items indicated the patient's perception that the staff member had been more caring. The instrument had already been tested psychometrically (DiNapoli, Turkel, Nelson, & Watson, 2010). It was also tested on 370 patients prior to testing the models outlined in Chapter 10.

Findings were similar to DiNapoli et al. (2010), with all 10 items loading as a single construct of caring with good reliability. The 10 behaviors, as reported by Nelson, DiNapoli, Turkel, and Watson (2011), are as follows:

1) Cultivating the practice of loving kindness and equanimity toward self and others. Loving kindness includes listening to, respecting, and identifying vulnerabilities in self and others.
2) Being authentically present: enabling, sustaining, and honoring faith and hope which is future-oriented and includes self-discovery.
3) Cultivating one's own spiritual practices and transpersonal self, going beyond ego-self.
4) Developing and sustaining a helping-trusting caring relationship.
5) Being present to, and supportive of, the expression of positive and negative feelings.
6) Creative use of self and all ways of knowing as part of the caring process; engaging in the artistry of Caritas (caring). At the core here is creative problem solving.
7) Engaging in genuine teaching-learning experience that attends to unity of being and subjective meaning: attempting to stay within others' frame.
8) Creating a healing environment at all levels.
9) Administering sacred acts of caring-healing by tending to basic needs.
10) Opening and attending to spiritual/mysterious and existential unknowns of life-death. This is belief in the impossible (miracles), even when others may assert doubt.[1]

Tool 2: The Caring Professional Scale (CPS)

The Caring Professional Scale (CPS) (Swanson, 2008) was developed to measure the patient's perception of the five caring behaviors in Swanson's Theory of Caring Behaviors (1999). The CPS had 18 items which used a 5-point Likert scale, asking whether the behavior was demonstrated by the provider of care, with the number (1) indicating that the behavior was "definitely" demonstrated and (5) indicating "no, not at all." In our

---

1 Nelson et al., 2011.

study, the scores were reversed for easier communication of the results, so that (1) indicated that the behavior was not demonstrated and (5) indicated the behavior was demonstrated. The 18 Caring Professional Scale items were summed for a possible range of 18–90 points. There were three negatively phrased items, including one asking whether the provider of care was abrupt, insulting, and emotionally distant. The 18-item CPS was examined on 324 patients from this study prior to using it for examining relationships with other measures. In a factor analysis, the three negatively phrased items loaded together, separately from the positively phrased items. A second factor analysis was run with just the 15 positively phrased items which loaded as 2 subscales: 1 with 10 items that addressed skills patients believed demonstrated relational or social competence, which we termed "compassionate healer items," and 5 items patients believed demonstrated technical competence, which we termed "competent practitioner items." The 15 items had a better model fit and were intuitively easier to explain from a caring perspective. It was decided to use the 15 positively phrased items with a total score possible of 15–75. Each subscale was titled for purposes of discussion of subscales found. Items of the CPS are listed in Table K.1.

Table K.1 CPS items and subscales from factor analysis.

Compassionate Healer Items (factor loading)	Competent Practitioner Items (factor loading)
Offers patient hope (.957)	Clinically competent (.781)
An attentive listener (.861)	Understanding (.695)
Centered on patient experience (.810)	Positive (.626)
Touched by patient experience (.801)	Technically skilled (.466)
Supportive (.721)	Personal (.449)
Comforting (.674)	
Caring (.661)	
Respectful of patient (.516)	
Informative (.513)	
Aware of patient's feelings (.493)	

Tool 3: The Caring Assessment Tool (CAT)

The Caring Assessment Tool (CAT) was used in our study for those hospitals implementing the Caring Model by Duffy. This tool measures patients' perception of caring by nurses, based on Watson's Human Caring Theory from 1979 and 1985 (Duffy, Brewer, & Wheaton, 2014). Psychometrics of the 27-item CAT tool used a 5-point Likert scale with higher scores indicating that the clinician had been more caring. Psychometrics of the CAT are described in Duffy et al. (2014). The items are summed for a possible score ranging from 27 to 135 points. It was not possible for us to assess the psychometrics in this study due to only 40 patients using the CAT in only one of the participating hospitals. Thus, the literature was relied upon for validity and reliability of the tool. Duffy et al. (2014) found that the 27 items loaded as a single factor with all items loading to show that each item contributed to the measuring of the construct of human caring. Duffy et al. (2014) provided an example of nine of the items from the CAT factor analysis. Each item is used to complete the statement "Since I have been a patient here, the nurses . . . ." Sample items to complete the statement from the CAT have been provided by Duffy et al.:

Since I have been a patient here, the nurses. . .

- . . .help me feel less worried.
- . . .anticipate my needs.
- . . .are responsive to my family.
- . . .help me explore different ways of dealing with my health problems.
- . . .treat my body carefully.
- . . .pay attention to me when I am talking.
- . . .help me see some good aspects of my situation.
- . . .make me feel as comfortable as possible.
- . . .help me understand how I am thinking about my illness.
  (2014, p. 88)

Three Tools for Assessing "Caring for the Patient" as Perceived By Staff Members

There were three tools available to measure staff members' perceptions of their own caring behaviors for patients.

Tool 1: The Caring Factor Survey—Care Provider Version (CFS-CPV)

The Caring Factor Survey—Care Provider Version (CFS-CPV) was derived from the CFS (Johnson, 2011) and has been shown to be a psychometrically sound tool to measure caring (Nelson, Thiel, Hozak, & Thomas, 2016). A factor analysis of 325 respondents from this study, prior to analysis of the model for this study, revealed it to be a valid and reliable tool to measure Watson's processes of Caritas (caring) for patients as perceived by the staff.

Tool 2: The Caring Professional Scale—Care Provider Version (CPS-CPV)

The Caring Professional Scale—Care Provider Version (CPS-CPV) is a derivation work premised on Swanson's Theory of Caring Behaviors (2008) and designed for providers of care to indicate how closely they perceive themselves to be providing caring behaviors as proposed by Swanson. There were 482 staff members from two hospitals who responded to every item of the CPS-CPV, and their responses were used for factor analysis. The items loaded as two subscales, just as they did in the CPS measuring patient perceptions of care, but it was found that the questions assessing "competent practitioner" contained only two items ("clinically competent" and "technically skilled") and that all 13 of the remaining items were seen by these staff members as assessing "compassionate healer." This was interesting to the study team because it indicated that while patients equated items such as "understanding," "positive," and "personal" (see Table K.1) with overall clinical competence, clinicians limited their definition of the competent practitioner to having "clinical competence" and being "technically skilled." This suggests that patients consider a competent practitioner as one who has relational skills as well as technical skills, and, quite possibly, that members of the healthcare team could see a practitioner as competent even in the absence of adequate relational skills. Items and factor loadings for the CPS-CPV are noted in Table K.2.

Tool 3: The Caring Efficacy Scale (CES)

The Caring Efficacy Scale (CES) was premised on Watson's Theory of Transpersonal Human Caring and Bandura's Self-efficacy Theory (1977). It contains 30 items that use a 6-point

Table K.2 CPS-CPV items and subscales from factor analysis.

Compassionate Healer Items (factor loading)	Competent Practitioner Items (factor loading)
Offers patient hope (.957)	Clinically competent (.781)
An attentive listener (.861)	Understanding (.695)
Centered on patient experience (.810)	Positive (.626)
Touched by patient experience (.801)	Technically skilled (.466)
Supportive (.721)	Personal (.449)
Comforting (.674)	
Caring (.661)	
Respectful of patient (.516)	
Informative (.513)	
Aware of patient's feelings (.493)	

Likert scale, which ranges from strongly disagree (1) to strongly agree (6). Psychometrics of the tool have been described elsewhere and reported it to be a valid and reliable tool (Coates, 2019). Only 52 respondents in this study used the CES; thus factor analysis was not possible, so psychometric information as described by Coates (2019) was relied upon for validity and reliability.

Tools to Measure Self-Care; Nurse Job Satisfaction; and Clarity of Self, Role, and System

Three additional tools were used to measure variables suspected of being likely predictors of caring as perceived by both patients and those providing care.

The Tool for Assessing Caring for Self as Perceived by the Staff

The Caring Factor Survey—Caring for Self (CFS-CS) is a derivative of the Caring Factor Survey (Lawrence & Kear, 2011) and was the only tool available for measuring caring for self. This is a limitation of the study, as not all of the people we assessed on caring for self were using Watson's theory. However, it did measure the concept of caring for self, and several items were similar to the measures for caring for patients; so, it was deemed sufficient to capture the construct of caring for self, no matter what theory was used. In addition, most of the caring theories used for this study were premised on or derived from Watson's Theory of Transpersonal Caring (1979, 1985, 2008b). The CFS-CS has been used in nursing in several countries (Itzhaki et al., 2015; Nelson, Itzhaki, et al., 2011) and was deemed appropriate for this study as well. The 10-item CFS-CS uses a 7-point Likert scale with higher scores indicating more caring for self. The CFS-CS was tested psychometrically using 856 respondents from the sample in this study prior to analysis of the model. Results revealed it to be a valid and reliable tool.

The Tool for Assessing Nurse Job Satisfaction

To measure job satisfaction, the Healthcare Environment Survey (HES) was utilized. The 59-item HES was used to assess satisfaction with the social and technical dimensions of the work environment. Dimensions of the HES are noted here:

The four relational dimensions of nursing are:

1) Patient care delivery
2) Relationship with physicians
3) Relationship with co-workers (nurses and other co-workers)
4) Relationship with unit manager

The seven technical dimensions of nursing are:

1) Autonomy
2) Workload
3) Professional growth
4) Executive leadership
5) Organizational rewards (including pay)
6) Staffing
7) Resources

A 7-point Likert scale was used, which ranges from strongly disagree (1) to strongly agree (7) with higher scores indicating greater job satisfaction. The HES has been found to be a psycho-metrically sound tool (Nelson, Persky, et al., 2015).

The Tool for Assessing "Clarity of Self, Role, and System"

Felgen's Measure of Clarity was developed by Felgen and Nelson (2016). Clarity is an important multidimensional construct that better enables nurses and other care providers to carry out their professional role. Here are Felgen's definitions for clarity of self, role, and system:

- Clarity of self includes self-awareness, emotional maturity, and purposefulness.
- Clarity of role relates to understanding expected goals, boundaries, and functions within one's role; specifically, the technical, relational, and innovative elements of one's professional role.
- Clarity of system involves staff members' understanding of how the professional role fits within an infrastructure of empowerment; how it is embedded within the organization's strategic, operational, and functional aspects; and their sense of self and understanding of who they are within the organization.[2]

This tool was psychometrically tested on 161 of the respondents in the second phase of this study, prior to testing of Model 2, and was found to be a valid and reliable tool.

---

2 Felgen & Nelson, 2016.

Appendix L

Data From Pause and Flow Study Related to Participants' Ability to Recall Moments of Pause and Flow Easily or with Reflection

In the pause and flow study reported in Chapter 11 of this book, here is the data related to how easily pause and flow were remembered.

Data on Participants' Ability to Recall Pause

When asked whether the most considerable pause was noticed right away or needed to be thought about through reflecting about the day, 289 people responded to the question (94%). Of the 289 who responded, 258 reported that the pause was immediately recalled (89%) while 31 reported having to reflect (11%).

When asked whether the second most considerable pause was recalled immediately or needed to be recalled through reflection, 279 people responded to the question (91%). Of the 279 who responded, 229 reported that the pause was immediately recalled (82%) while 50 reported having to reflect (18%).

When asked whether the third most considerable pause was recalled immediately or needed to be recalled through reflection, 249 people responded to the question (81%). Of the 249 who responded, 160 reported that the pause was immediately recalled (64%), while 89 reported having to reflect (36%).

Data on Participants' Ability to Recall Flow

When asked whether the most considerable moment of flow was recalled immediately or recalled through reflection, 283 people responded to the question (92%). Of the 283 who responded, 222 reported that a moment of flow was recalled immediately (78%), while 61 reported that a moment of flow could only be recalled through reflection (22%).

When asked whether the second most considerable moment of flow was recalled immediately or recalled through reflection, 272 people responded to the question (86%). Of the 272 who responded, 195 reported that their second most considerable moment of flow was recalled immediately (72%), while 77 reported that it was recalled through reflection (28%).

When asked whether the third most considerable moment of flow was recalled immediately or recalled through reflection, 239 people responded to the question (78%). Of the 239 who responded, 151 reported that it was immediately recalled (63%) while 88 reported that it was recalled through reflection (37%).

As was indicated in Chapter 11, our aim in asking about which instances could more easily be recalled was to determine which instances—pause or flow—were more impactful. We discovered that moments of pause were remembered more often and more easily than moments of flow. While they were more easily remembered, more study is required to fully understand their impact.

Appendix M

Identified Pauses and Proposed Interventions Resulting from a Pause and Flow Study

Three issues identified by the staff of a 650-bed urban tertiary hospital in the Northeastern United States were:

1) Pharmacy not dispensing patient's medications on time or at all,
2) Telephone interruptions of the nurses, and
3) Lack of functional equipment (BP machines, O_2 regulators, bed alarms).

The interventions proposed to solve these pauses are explained below.

Pause

Pharmacy not dispensing patient's medications on time or at all.

Intervention

The Director of Pharmacy became a partner in resolving this issue. Satellite pharmacy stations were implemented on every unit to expedite medication communication and dispensing. This provided time to talk with pharmacy staff members if questions arose regarding dispensing of the medication or regarding the medication itself. This intervention allowed nurses to continue with patient care with fewer interruptions, better communication with patients, and timelier dispensing of medications. Through this intervention, instances of pause, due to issues with medications not being dispensed, were reduced.

Pause

Telephone interruptions of the nurses.

Intervention

A Unit Associate (similar to a Health Unit Coordinator) would take the message/phone numbers of nonemergency calls and inform the caller that the nurse would get back to them as soon as possible. After taking the call, the Unit Associate would give the appropriate phone number to the nurse to return the phone call. Any urgent calls would be transferred to the resource nurse/nurse immediately. After implementation, management of family calls improved, and quicker attention was paid to true emergencies, as it was identified that an interruption of the nurse now indicated urgency for the nurse to take the call immediately.

Pause

Lack of functional equipment (BP machines, O_2 regulators, bed alarms).

Intervention

Three BP machines were ordered to replace the broken ones and ensure that there were always enough blood pressure monitoring machines. Six oxygen regulators were ordered, as were seven more bed alarms with chair and bed pads. This reduced waiting for equipment and enhanced monitoring of patients to enhance patient safety.

Appendix N

Factors Related to a Focus on Pain Versus Factors Related to a Focus on Comfort

Table N.1 Comparison of pain and comfort.

Pain focus	Comfort focus
Pain is a detractor of comfort (Hamilton, 1989).	Comfort is a positive modulator of pain.
Pain has a connotation of weakness and vulnerability.	Comfort has a connotation of strengthening, being all you can be, and hope (Kolcaba, 1991).
Singular	Holistic
Usually excludes assessment of discomfort external to physiological sensation.	Includes assessment of absence or presence of pain and essence of well-being.
Excludes exploring or believing the patient's story.	Includes exploring and believing the patient's story.
Isolated—pain focus.	Integrative—patient focus.
Pain is usually associated with organ, injury, trauma or disease (headache, stomachache, chest pain).	Patient determines location and sources of discomfort (lost, hopeless, achy, cannot eat or sleep), and includes examples under pain.
Domination Paradigm, which includes four core components: hierarchical domination, ranking of men over women, culturally accepted abuse and violence, and belief of domination and submission is inevitable (Eisler & Potter, 2014).	Partnership Paradigm, which includes four core components: egalitarian structure, equal partnership of men and women, abuse and violence is not accepted, and support of empathic and mutually respectful relations (Eisler & Potter, 2014).
Goal is limited to pain management and may make the patient more uncomfortable.	Goal encompasses, enhances, and integrates pain management into the whole of care.

Appendix O

Comfort/Pain Perception Survey (CPPS)—Patient Version

Comfort/Pain Perception Survey (CPPS) - Patient

Using the 7-point scale, please select your degree of agreement with each statement. The more strongly you disagree, the more to the left of number 4 you should select. The more you agree with the statement, the more to the right of the number 4 you should select. For example, if you strongly disagree with the statement, select number 1. If you strongly agree with the statement, select number 7. If you feel neutral about the statement, select number 4, that you neither disagree or agree.

Strongly Disagree ①	Disagree ②	Slightly Disagree ③	Neutral ④	Slightly Agree ⑤	Agree ⑥	Strongly Agree ⑦

1. I have a sense of well-being.

1○　2○　3○　4○　5○　6○　7○

2. I have no pain.

1○　2○　3○　4○　5○　6○　7○

3. I have no fear.

1○　2○　3○　4○　5○　6○　7○

4. I am not suffering.

1○　2○　3○　4○　5○　6○　7○

5. I have trust in the healthcare system.

1○　2○　3○　4○　5○　6○　7○

6. I have experienced a loss of physical functioning because of my pain (select yes or no).

    ○ Yes
    ○ No (if selecting no, skip items 7 and 8 below and continue with item 9)

Figure O.1 Part 1　Comfort/pain perception (CPPS) survey—for patient.

7. I have grief related to loss of physical function that occurred because of my pain.
Note: Only respond to this item if you responded "yes" to item 6 above.

1 O   2 O   3 O   4 O   5 O   6 O   7 O

8. I have accepted the loss of physical function that occurred because of my pain.
Note: Only respond to this item if you responded "yes" to item 6 above.

1 O   2 O   3 O   4 O   5 O   6 O   7 O

9. My care team here excludes me in planning my care.

1 O   2 O   3 O   4 O   5 O   6 O   7 O

10. I do not have a sense of well-being

1 O   2 O   3 O   4 O   5 O   6 O   7 O

11. I am experiencing pain.

1 O   2 O   3 O   4 O   5 O   6 O   7 O

12. I have fear.

1 O   2 O   3 O   4 O   5 O   6 O   7 O

13. I am suffering.

1 O   2 O   3 O   4 O   5 O   6 O   7 O

14. I have no trust in the healthcare system.

1 O   2 O   3 O   4 O   5 O   6 O   7 O

15. I feel worry.

1 O   2 O   3 O   4 O   5 O   6 O   7 O

Figure O.1 Part 2   Continued

16. I feel bitterness.

1 ○ 2 ○ 3 ○ 4 ○ 5 ○ 6 ○ 7 ○

17. I feel anger.

1 ○ 2 ○ 3 ○ 4 ○ 5 ○ 6 ○ 7 ○

18. I feel helpless.

1 ○ 2 ○ 3 ○ 4 ○ 5 ○ 6 ○ 7 ○

19. I feel hopeless.

1 ○ 2 ○ 3 ○ 4 ○ 5 ○ 6 ○ 7 ○

20. I feel lonely.

1 ○ 2 ○ 3 ○ 4 ○ 5 ○ 6 ○ 7 ○

21. I feel I have little value.

1 ○ 2 ○ 3 ○ 4 ○ 5 ○ 6 ○ 7 ○

22. Everyday I am here, I see that the care is provided with loving kindness.

1 ○ 2 ○ 3 ○ 4 ○ 5 ○ 6 ○ 7 ○

23. As a team, my caregivers are good at creative problem solving to meet individual needs and requests.

1 ○ 2 ○ 3 ○ 4 ○ 5 ○ 6 ○ 7 ○

24. The care providers honoured my own faith, helped in still hope, and respected my belief system as part of my care.

1 ○ 2 ○ 3 ○ 4 ○ 5 ○ 6 ○ 7 ○

Figure O.1 Part 3   Continued

25. When my caregivers teach me something new, they teach me in a way that I can understand.

1○ 2○ 3○ 4○ 5○ 6○ 7○

26. My caregivers encouraged me to practice my own individual spiritual beliefs of my self-caring and healing.

1○ 2○ 3○ 4○ 5○ 6○ 7○

27. My caregivers have responded to me as a whole person, helping to take care of all my needs and concerns.

1○ 2○ 3○ 4○ 5○ 6○ 7○

28. My caregivers have established a helping and trusting relationship with me during my time here.

1○ 2○ 3○ 4○ 5○ 6○ 7○

29. My healthcare team has created a healing environment that recognizes the connection between my body, mind, and spirit.

1○ 2○ 3○ 4○ 5○ 6○ 7○

30. I feel like I can talk openly and honestly about what I am thinking, because those who are caring for me embrace my feelings, no matter what my feelings are.

1○ 2○ 3○ 4○ 5○ 6○ 7○

31. My caregivers are accepting and supportive of my beliefs regarding a higher power, which allows for the possibility of me and my family to heal.

1○ 2○ 3○ 4○ 5○ 6○ 7○

32. Those who are caring for me believe I am in pain when I tell them I am in pain.

1○ 2○ 3○ 4○ 5○ 6○ 7○

Figure O.1 Part 4   Continued

33. Those who are caring for me understand my loss as it relates to my discomfort and pain.

1 ○   2 ○   3 ○   4 ○   5 ○   6 ○   7 ○

34. The interprofessional team is working together to improve my comfort.

1 ○   2 ○   3 ○   4 ○   5 ○   6 ○   7 ○

35. I have discomfort that is not being addressed.

1 ○   2 ○   3 ○   4 ○   5 ○   6 ○   7 ○

36. I am comfortable.

1 ○   2 ○   3 ○   4 ○   5 ○   6 ○   7 ○

37. I need relief from my discomfort.

1 ○   2 ○   3 ○   4 ○   5 ○   6 ○   7 ○

Figure O.1 Part 5   Continued

Appendix P

Comfort/Pain Perception Survey (CPPS)—Care Provider Version

Care Provider Beliefs about Pain and Comfort

Using the 7-point scale, please select your degree of agreement with each statement. The more strongly you disagree, the more to the left of number 4 you should select. The more you agree with the statement, the more to the right of the number 4 you should select, For example, if you strongly disagree with the statement, select number 1. If you strongly agree with the statement select number 7, if you feel neutral about the statement select number 4, that you neither disagree or agree.

Strongly Disagree	Disagree	Slightly Disagree	Neutral	Slightly Agree	Agree	Strongly Agree
①	②	③	④	⑤	⑥	⑦

1. My primary intervention for treating pain is pain medication.

1○ 2○ 3○ 4○ 5○ 6○ 7○

2. My primary intervention for treating pain is non-medicine or non-Opioid interventions.

1○ 2○ 3○ 4○ 5○ 6○ 7○

3. I believe the interventions I provide to improve the patients comfort can increase comfort beyond what pain medications do.

1○ 2○ 3○ 4○ 5○ 6○ 7○

4. I believe the interventions that I provide to improve the patients comfort can help when pain medications do not.

1○ 2○ 3○ 4○ 5○ 6○ 7○

5. When I provide pain medication(s) or intervention(s) to improve comfort, I think about the unique needs of the patient that may be contributing to the pain and not only the pain itself.

1○ 2○ 3○ 4○ 5○ 6○ 7○

Figure P.1 Part 1   Comfort/pain perception (CPPS) survey—for staff.

6. When I administer pain medication(s) or intervention(s) to improve the patients comfort, I do so to be in alignment with the patient's spirituality.

1○ 2○ 3○ 4○ 5○ 6○ 7○

7. When I administer pain medication(s) or intervention(s) to improve the patients comfort, I do so to be in alignment with the patient's values.

1○ 2○ 3○ 4○ 5○ 6○ 7○

8. When I administer pain medication(s) or intervention(s) to improve the patients comfort, I do so to be in alignment with the patient's beliefs.

1○ 2○ 3○ 4○ 5○ 6○ 7○

9. I believe being curious about a patient helps me ask proper questions and/or interact with them in such a way to understand patient better.

1○ 2○ 3○ 4○ 5○ 6○ 7○

10. When I provide care to a patient, I try to ask questions or interact with them in such a way that helps me understand who that person is.

1○ 2○ 3○ 4○ 5○ 6○ 7○

11. I believe I am able to leave all of my own bias and prejudice behind me when I provide care to a patient.

1○ 2○ 3○ 4○ 5○ 6○ 7○

12. I believe all patients are unique.

1○ 2○ 3○ 4○ 5○ 6○ 7○

13. I enjoy knowing the patients I provide care to.

1○ 2○ 3○ 4○ 5○ 6○ 7○

Figure P.1 Part 2   Continued

14. I believe the more I learn about the patients I provide care to, the greater the comfort I am able to provide.

1○ 2○ 3○ 4○ 5○ 6○ 7○

15. As the relationship with my patient improves, my ability to provide good pain management improves.

1○ 2○ 3○ 4○ 5○ 6○ 7○

16. I believe patients feel safer when they feel they are understood by their care providers.

1○ 2○ 3○ 4○ 5○ 6○ 7○

17. Patients want me to understand their unique needs related to pain management.

1○ 2○ 3○ 4○ 5○ 6○ 7○

18. Patients want me to understand their unique needs related to comfort.

1○ 2○ 3○ 4○ 5○ 6○ 7○

19. It is important for me to be aware of the patient needing adjustments in pain management while the patient is within my care.

1○ 2○ 3○ 4○ 5○ 6○ 7○

Figure P.1 Part 3   Continued

Satisfaction with Relationship with Coworkers

Strongly Disagree	Disagree	Slightly Disagree	Neutral	Slightly Agree	Agree	Strongly Agree
①	②	③	④	⑤	⑥	⑦

20. I am satisfied with how easy it is for new employees to feel welcome in my unit or department.

1○ 2○ 3○ 4○ 5○ 6○ 7○

21. I am satisfied with the teamwork and cooperation in the unit/department I work in.

1○ 2○ 3○ 4○ 5○ 6○ 7○

22. I am satisfied with how friendly and outgoing the people on my unit or department are.

1○ 2○ 3○ 4○ 5○ 6○ 7○

23. I am satisfied with how people I work with on my unit or department get along, no matter what the level of their education and experience is.

1○ 2○ 3○ 4○ 5○ 6○ 7○

24. I am satisfied with how people I work with in my unit/department help me out when I get really busy and need help.

1○ 2○ 3○ 4○ 5○ 6○ 7○

25. I am satisfied with how nurses at the hospital/facility show respect for other staff members.

1○ 2○ 3○ 4○ 5○ 6○ 7○

Figure P.1 Part 4 Continued

26. I am satisfied with how nurses in general cooperate with other staff in my unit or department.

1○ 2○ 3○ 4○ 5○ 6○ 7○

27. I am satisfied with how well nurses and other staff in my unit/department work together as a team.

1○ 2○ 3○ 4○ 5○ 6○ 7○

28. I am satisfied with how nurses I work with are respectful of the skill and knowledge of all the staff in my unit/department.

1○ 2○ 3○ 4○ 5○ 6○ 7○

29. I am satisfied with how nurses at this hospital/facility generally understand and appreciate what all other staff members do.

1○ 2○ 3○ 4○ 5○ 6○ 7○

30. I am satisfied with how physicians general cooperate with staff in my unit or department.

1○ 2○ 3○ 4○ 5○ 6○ 7○

31. I am satisfied with how physicians and staff in my unit/department work together as a team.

1○ 2○ 3○ 4○ 5○ 6○ 7○

32. I am satisfied with how physicians I work with are respectful of the skill and knowledge of all the staff in my unit/department

1○ 2○ 3○ 4○ 5○ 6○ 7○

33. I am satisfied with how physicians at this hospital/facility generally understand and appreciate what all the staff members do.

1○ 2○ 3○ 4○ 5○ 6○ 7○

34. I am satisfied with how physicians at this hospital/facility show respect for staff members.

1○ 2○ 3○ 4○ 5○ 6○ 7○

Figure P.1 Part 5   Continued

Clarity of Role

Strongly Disagree	Disagree	Slightly Disagree	Neutral	Slightly Agree	Agree	Strongly Agree
①	②	③	④	⑤	⑥	⑦

35. I have clear planned goals and objectives for my job.

1○ 2○ 3○ 4○ 5○ 6○ 7○

36. I know that I have divided my time properly.

1○ 2○ 3○ 4○ 5○ 6○ 7○

37. I know what my responsibilities are.

1○ 2○ 3○ 4○ 5○ 6○ 7○

38. I know exactly what is expected of me.

1○ 2○ 3○ 4○ 5○ 6○ 7○

39. Explanation is clear for me what has to be done.

1○ 2○ 3○ 4○ 5○ 6○ 7○

Figure P.1 Part 6   Continued

Clarity of System

Strongly Disagree ①	Disagree ②	Slightly Disagree ③	Neutral ④	Slightly Agree ⑤	Agree ⑥	Strongly Agree ⑦

40. I understand how patient assignments are made as it relates to continuity of care.

1○ 2○ 3○ 4○ 5○ 6○ 7○

41. I understand how patient assignments are made as it relates to hospital policy.

1○ 2○ 3○ 4○ 5○ 6○ 7○

42. I understand how schedules are made including how part-time and full-time staff are assigned.

1○ 2○ 3○ 4○ 5○ 6○ 7○

43. I understand how the schedule is made in consideration of vacation, education classes for staff, and other necessary scheduling requirements for staff.

1○ 2○ 3○ 4○ 5○ 6○ 7○

44. I understand what our organization's key success is and how it makes us stand apart from other hospitals\facilities.

1○ 2○ 3○ 4○ 5○ 6○ 7○

45. I understand the difference between responsibility, authority and accountability.

1○ 2○ 3○ 4○ 5○ 6○ 7○

46. I understand practice change (what I do in my job) is linked to principle (a rationale or reason).

1○ 2○ 3○ 4○ 5○ 6○ 7○

47. I believe in shared governance where staff and managers both have input into decisions.

1○ 2○ 3○ 4○ 5○ 6○ 7○

48. I believe managers should support staff so staff can manage patients.

1○ 2○ 3○ 4○ 5○ 6○ 7○

49. I believe unit practice controls (small group of unit staff leaders) are helpful in setting unit policy and helping make unit decisions.

1○ 2○ 3○ 4○ 5○ 6○ 7○

Figure P.1 Part 7 Continued

Clarity of Self

Strongly Disagree ①	Disagree ②	Slightly Disagree ③	Neutral ④	Slightly Agree ⑤	Agree ⑥	Strongly Agree ⑦

50. I know who I am.

1○ 2○ 3○ 4○ 5○ 6○ 7○

51. I know my strengths.

1○ 2○ 3○ 4○ 5○ 6○ 7○

52. I know my weaknesses.

1○ 2○ 3○ 4○ 5○ 6○ 7○

53. I know what brings me joy.

1○ 2○ 3○ 4○ 5○ 6○ 7○

54. I understand what my purpose in life is.

1○ 2○ 3○ 4○ 5○ 6○ 7○

55. I understand how to use myself to help others.

1○ 2○ 3○ 4○ 5○ 6○ 7○

Figure P.1 Part 8 Continued

56. I am mindful of how I behave.

1○ 2○ 3○ 4○ 5○ 6○ 7○

57. I am mindful of how I behave in relation to how my behavior impacts others.

1○ 2○ 3○ 4○ 5○ 6○ 7○

58. I am mindful of how I speak.

1○ 2○ 3○ 4○ 5○ 6○ 7○

59. I am mindful of how I speak in relation to how my speech impacts others.

1○ 2○ 3○ 4○ 5○ 6○ 7○

60. I recognize that I need to take time for self care because I need to replenish my own energy.

1○ 2○ 3○ 4○ 5○ 6○ 7○

61. I believe self care is about me replenishing my energy.

1○ 2○ 3○ 4○ 5○ 6○ 7○

62. I believe self care is about me replenishing my energy so I can serve others better.

1○ 2○ 3○ 4○ 5○ 6○ 7○

63. I am aware of own prejudice and bias.

1○ 2○ 3○ 4○ 5○ 6○ 7○

64. I am intentional about the words I use to speak with others.

1○ 2○ 3○ 4○ 5○ 6○ 7○

Figure P.1 Part 9 Continued

65. I am intentional about my behavior when I interact with others.

1○ 2○ 3○ 4○ 5○ 6○ 7○

66. I am the only person who is responsible for how I speak.

1○ 2○ 3○ 4○ 5○ 6○ 7○

67. I am the only person who is responsible for how I behave.

1○ 2○ 3○ 4○ 5○ 6○ 7○

68. I take time to think about who I am.

1○ 2○ 3○ 4○ 5○ 6○ 7○

69. I take time to reflect on my purpose in life.

1○ 2○ 3○ 4○ 5○ 6○ 7○

70. I take time to reflect on how my speech impacts others.

1○ 2○ 3○ 4○ 5○ 6○ 7○

71. I take time to reflect on how my behavior impacts others.

1○ 2○ 3○ 4○ 5○ 6○ 7○

72. When I come to understand how I might help others better, I adjust how I speak.

1○ 2○ 3○ 4○ 5○ 6○ 7○

73. When I come to understand how I might help others better, I adjust how I behave.

1○ 2○ 3○ 4○ 5○ 6○ 7○

Figure P.1 Part 10   Continued

74. I am in change of my life

1 ○ 2 ○ 3 ○ 4 ○ 5 ○ 6 ○ 7 ○

75. I own my own behaviors.

1 ○ 2 ○ 3 ○ 4 ○ 5 ○ 6 ○ 7 ○

76. I love who I am

1 ○ 2 ○ 3 ○ 4 ○ 5 ○ 6 ○ 7 ○

77. I enjoy using who I am to help serve others.

1 ○ 2 ○ 3 ○ 4 ○ 5 ○ 6 ○ 7 ○

78. I use who I am to serve others.

1 ○ 2 ○ 3 ○ 4 ○ 5 ○ 6 ○ 7 ○

79. I seek to do more of what brings me joy in life.

1 ○ 2 ○ 3 ○ 4 ○ 5 ○ 6 ○ 7 ○

80. I understand what I do not like in life.

1 ○ 2 ○ 3 ○ 4 ○ 5 ○ 6 ○ 7 ○

81. For those things in life that I do not enjoy, or even dislike. I understand why.

1 ○ 2 ○ 3 ○ 4 ○ 5 ○ 6 ○ 7 ○

82. For those things in life that I enjoy, I understand why.

1 ○ 2 ○ 3 ○ 4 ○ 5 ○ 6 ○ 7 ○

83. I pay attention to what I am good at.

1 ○ 2 ○ 3 ○ 4 ○ 5 ○ 6 ○ 7 ○

84. I pay attention to what I am not good at.

1 ○ 2 ○ 3 ○ 4 ○ 5 ○ 6 ○ 7 ○

Figure P.1 Part 11 Continued

Appendix Q

Predictors of OUD

Table Q.1 Risk factors for OUD.

	Ciesielski et al. (2016)	Cochran et al. (2014)	Zamirinejad, Hojjat, Moslem, Moghaddam Hosseini and Akaberi (2018)
Demographics			
Younger age	X	X	
Gender (male)	X	X	
Employee (vs. dependent/spouse)		X	
Region of US, South	X		
Region of US, West	X		
Region of US, Midwest	X		
Psychiatric			
Greater rates of psychiatric disorders		X	
Mental illness	X		
Early maladaptive schemas			
Emotional depravity			X
Abandonment/ instability			X

Table Q.1 (Continued)

	Ciesielski et al. (2016)	Cochran et al. (2014)	Zamirinejad, Hojjat, Moslem, Moghaddam Hosseini and Akaberi (2018)
Mistrust/abuse			X
Social isolation/ alienation			X
Defectiveness/shame			X
Failure to achieve			X
Dependence/ incompetence			X
Vulnerability to harm			X
Enmeshment/ undeveloped self			X
Subjugation of needs			X
Emotional inhibition			X
Insufficient self-control			X
Entitlement/grandiosity			X
Unrelenting standards			X
Drug and alcohol use			
Drug prescriptions filled at more pharmacies	X	X	
Drug prescriptions filled by multiple prescribers	X		
Prescription history of more opioids		X	
More days supply of opioids		X	
30-day prescriptions	X		
Prescribed more concomitant medications		X	
Non-opioid substance abuse	X		
Chronic opioid user	X		

(*Continued*)

Table Q.1   (Continued)

	Ciesielski et al. (2016)	Cochran et al. (2014)	Zamirinejad, Hojjat, Moslem, Moghaddam Hosseini and Akaberi (2018)
Daily opioid use (≥120 mg/d morphine equivalent)	X		
Nondependent alcohol abuse	X		
Tobacco use disorder	X		
Use of medical services			
Utilize more medical and psychiatric services		X	

Appendix R

Personal Qualities of Clinicians and Others Suited to Become Trusted Others

Some personal qualities and therapeutic practices that may help clinicians become the kinds of trusted others who can be of effective long-term service to people in recovery are (i) modeling "not knowing," (ii) cultivating self-awareness, and (iii) committing to seeking the perspectives of others.

Modeling "Not Knowing"

While people often enter treatment feeling starved for answers, it may be ideal for clinicians working with people in recovery to facilitate a "partnership paradigm" rather than a "domination paradigm" (Eisler & Potter, 2014). It is the position of the authors of this chapter that telling people in recovery what to do and how to do it is of less benefit in the long term than positioning one's self as a peer (or partner) on the same journey. Even if clinicians working with people in recovery have not had the same experiences with OUD or SUD, common ground can be found in the universal reality that all humans struggle at times to find a way forward, and perhaps most importantly, that none of us ever really knows exactly what to do in life. It is helpful, we believe, for clinicians to be honest about what they do not know, as this helps the people they are serving to join them in the very common human condition of not knowing.

This appreciation for the condition of not knowing serves to contrast the all-or-nothing thinking that is common in people in early recovery, while also acknowledging the very real need for a therapeutic partner to recognize that there are very specific contextual factors at play in the person's life. The key to discovering what helps or hinders long-term recovery for any individual is discovering what helps that person thrive. No preconceived notions about what helps people-in-general to thrive will be of much help in this quest. It is genuine curiosity on the part of both the person in recovery and the therapeutic partner that leads to discovering each person's unique path to wellness. Initially there may be huge discomfort for people in recovery with sitting in the unknown, but with practice, comfort grows.

When a person in recovery seeks an answer, rather than receiving an answer, the person may be better served by a clinician who presents an inquiry such as, "That's an important question. . .let's sit with it and see if we can figure something out." The job of the therapist is not to provide answers so much as it is to help the person learn to tolerate what is unknown and possibly unknowable on the way, at least sometimes, to finding some sort of provisional answer. When clinicians model not knowing, they are helping their clients and patients to rewire neural pathways (Hanson, 2018; Hani, 2017), making the quest for answers—and more importantly the occasional failure to find answers—a healthy part of self-discovery.

Cultivating Self-Awareness

If a therapist has thoughts that start, "the right way to do that is. . ., or the best way to do that is. . ." they may be working with an assimilationist mindset. In fact, if you are a therapist or counselor of any kind, a good general rule is to assume that some assimilationist thinking is active in you at all times, so that you can consistently question whether or not you are in fact engaging with someone in a way that fails to take into full consideration the exact context of their life—i.e. their race, gender, sexual orientation and/or identification, personal history, socioeconomic situation, etc. There is no one right way to live a life.

We submit that clinicians who are self-aware enough to individualize every therapeutic interaction are more likely to become trusted others.

We submit that clinicians who are self-aware enough to individualize every therapeutic interaction are more likely to become trusted others. If clinicians and the people they are treating are on a journey of self-awareness and self-discovery together, clinicians are far more likely to meet the needs of their patients and clients. Without a host of preconceived notions and pre-packaged wisdom clouding their perspectives, clinicians are more likely to help people in recovery discover, in any given moment, what they need most, right there and then. Every person in recovery is on a unique path, so clinicians who consider and address the uniqueness of each person, fully considering and integrating the realities of the person's context, will be able to provide customized interactions to match the person's life.

Commitment to Perspective-Seeking

We believe that one of the key opportunities to successfully follow people into long-term recovery is to ask whose perspectives and expertise may be missing from the team to make that happen. The process of pausing to notice who are the partners best suited to achieve successful outcomes all along the continuum of care sets a process in motion to notice, reflect, and integrate knowledge of who is needed on the team and what is needed at any point of the recovery process. It also has the benefit of modeling not knowing and the value of self-awareness, while also being a process individuals can continue to utilize throughout their recovery.

Within the process of exploring for missing partners, we propose exploring the impact of occupational therapy (OT) on the team. OT is one key discipline that is often not at the table where care is being designed or OT is simply not yet known or utilized to its fullest scope on the team due to not yet understanding the extent to which OT assists in improving the quality of one's life.

To more deeply explain this proposal for people who are unfamiliar with occupational therapy, OT practitioners are experts in understanding how a variety of factors such as occupational deprivation, disruption, and loss impact one's ability to engage and thrive in meaningful activities. The American Occupational Therapy Association has defined OT as "the only profession that helps people do the things they want and need to do through the therapeutic use of daily activities" (OATA, 2020, second paragraph).

Individual understanding for each patient is established through the OT's assessment of the interplay among individual, system, and environmental factors. This assessment facilitates understanding of how each individual's unique factors influence their function and subsequently informs what their next best steps are. Understanding of factors informing the patient's daily functioning is also helpful to improve team collaboration and problem-solving. During recovery, individuals are adapting to the loss or change of their occupations, roles, self-identity; they are adjusting or ceasing habits, routines, and rituals while simultaneously establishing new habits, routines, and rituals which support long-term recovery.

Appendix S

Qualities of Systems and Organizations Suited to Serve People Recovering from OUD

Organizations aiming to design systems that nurture clinicians to become trusted others should strive to embrace the same characteristics it wants to foster in its clinicians—modeling not knowing, cultivating self-awareness, and adopting an attitude of perspective seeking—but at the system level.

Advancing a Culture in Which "Not Knowing" Is Accepted

A culture in which it is unacceptable to be without answers is often a culture in which the status quo is embraced. It is also a culture in which innovation and growth—not just of individuals, but of the organization itself—will be stunted (Sinek, 2011). In the context of recovery, embracing the status quo often means using the traditional medical model and randomized clinical trials as the primary guides to care. One need look no further than the current opioid crisis to see an example of how devastating the results can be of listening to people driven by profit motives and allowing one's self to be guided by studies funded by profit-driven people and corporations (Bernard, Chelminski, Ives, & Ranapurwala, 2018).

In order to advance a culture in which "not knowing" is accepted, leaders must model the behavior themselves. When leaders at the highest level in an organization voice the reality that there are not enough perspectives at the table,

A culture in which it is unacceptable to be without answers is often a culture in which the status quo is embraced.

others will do the same, and ideas and innovations will flow (Sinek, 2017, 2019).

Advancing a Culture in Which Systemic Awareness Is Actively Sought

Systemic awareness is very much like self-awareness. Self-awareness requires looking deeply at ones' self, and systemic awareness requires looking deeply at the system in which one works or lives. A system comprises structures, processes, and people (Stroh, 2015). To assess a system, you look at the structures, processes, and policies in place in an organization and examine the effects they appear to be having on the people in the organization. Are the people in the organization willing to speak up, or do some of them seem reticent? Are people more apt to collaborate or to compete with each other? What structures and processes do you have in place that make it more likely not only for people to actively seek information from diverse perspectives but to do something with the information they find? How often is space made in your work culture for people to pause and reflect? Do the statements of the people with the most power ring true for the people with the least power? Seeking the answers to these and a host of similar questions will help an organization foster systemic awareness. Every structure, process, and policy has the potential to have unintended negative consequences. For example, if you put a structure in place that fosters competition, you will have stifled collaboration. Even if you find that consequence acceptable, you need to be aware of its potentially far-reaching implications. Systemic awareness, like self-awareness, is work that is ongoing, never complete.

Systemic awareness is not a task on a CEO's to-do list. It is the work of an entire organization (Stroh, 2015). What would happen in your organization if you introduced the notion of "systemic agreements," in order to develop organization-wide commitments that truly represent everyone in the organization? For example, an organization could have a systemic agreement to make policies that are antiracist—to look at each policy to see whether they propel people forward or hold them back. If a policy seems on first blush to propel people forward, does the policy, in fact, propel everyone equally? Is there anything about a given policy that shows "accidental assimilationist" attitudes? (This often shows up in culturally uninformed policies around bereavement.)

These aspects of systemic awareness are not just designed to foster a better culture within the organization. They are designed to foster better care and service for patients and clients as well. The systemically aware organization asks, "Who do we serve?" Then we ask ourselves how well our staff matches our clientele, and if there is a mismatch, what are we willing to do about it? It is not enough for individual clinicians to be culturally competent and aware of how easy it is to bring assimilationist attitudes to bear in one-to-one encounters. Organizations must also be culturally competent and aware of how easy it is to write assimilationist attitudes into policies (Kendi, 2017) and to design them into structures and processes.

Advancing a Culture in Which Perspective Seeking Is Prized

Organizations have the power to shut down perspective seeking or to actively foster it. It is not uncommon for organizations to design care based on the traditional medical model and the findings of randomized testing. If the current opioid crisis can teach us anything, it is that when these are the only perspectives driving care, patients and clients suffer. Some of the authors of this chapter were on the front lines of OUD care when the use of opioids became common in the United States, and it was apparent from very early on that opioids were far more addictive than the traditional medical model—and the pharmaceutical companies—admitted. In fact, in 2007, Purdue Pharma was fined $634.5 million for "knowingly disseminat[ing] false information" (Bernard, Chelminski, Ives, & Ranapurwala, 2018, p. 4) about the addictive properties of opioids. Before this landmark case, Purdue Pharma, the manufacturer of the opioids Percodan and Oxycontin, funded studies conducted by the America Pain Society and other seemingly reputable groups which yielded faulty findings (Bernard et al., 2018) which then guided the practice of countless medical professionals.

How might things have been different if front-line addiction counselors, patients, and their family members had been at the table when care was being designed?

How many organizations used the traditional medical model and randomized control trials to design care throughout the 1990s and 2000s? And how might things have been different if front-line addiction counselors, patients, and their family members had been at the table when care was being designed? What if the three-legged stool of evidenced-based practice—clinical experience, patient preferences, and clinical research—was used as a reflection tool for the effectiveness of approaches (Sackett, Rosenberg, Gray, Haynes, & Richardson, 1996)? If the observations of addiction professionals and patients and families reporting their first-hand experience of the addictiveness of these prescriptions was weighed as heavily as the evidenced-based practice model indicates it should be, would it have changed the outcome? What would happen if more time was given to truly listen to all the other voices?

In a culture in which perspective seeking is prized, here are some questions to ask:

- What are we not seeing as a board or as the group of people designing care?
- Do we have adequate diversity of role, race, sexual orientation, etc., in the group designing care? If not, what are we willing to do about it right now?
- Who can we invite in today, to tell us literally anything we don't already know?

Appendix T

Factor Loadings for Satisfaction with Staffing/Scheduling and Resources

Prior to our work in Jamaica, the subscales of "satisfaction with scheduling" and "satisfaction with resources" had not previously been part of the Healthcare Environment Survey (HES). As you can see from Figure T.1, in the Jamaica study we examined the factor loadings of the items related to satisfaction with scheduling and resources in an exploratory factor analysis. Findings revealed that all factor loadings were above .40, which meant the items were appropriate to include in the subscales of satisfaction with scheduling (items 59–62) and satisfaction with resources (items 63–65).

Item	Factor and Loadings		Explained Variance
	1	2	
59. I am satisfied with the amount of advance notice I have prior to my new shift roster starting.	.63		48.5%
60. I am satisfied with the shift rotation I am assigned.	.66		
61. I am satisfied with the input I have into my final schedule prior to the roster being posted.	.92		
62. I am satisfied with my ability to change my schedule after the schedule roster is posted, if I need to change it.	.75		
63. I am satisfied with the availability of supplies required to do my job.		.95	20.4%
64. I am satisfied with the availability of equipment needed to do my job.		.94	
65. I am satisfied with my access to clinical experts/specialists to do my job (may include physicians, pharmacists, clinical nurse specialists, etc.).		.41	

Figure T.1 Factor loadings for Satisfaction with Staffing/Scheduling and Satisfaction with Resources.

Appendix U

Detail Regarding Item Reduction of Instruments to Measure Caring

Both of the instruments—the 10-item instrument to measure caring for self and the 10-item instrument to measure the caring of the unit manager—retained five of six items that came verbatim from Watson's Theory of Transpersonal Caring (2008a). For example, the five items from Watson's work that were retained and used to measure caring for self were as follows:

1) I honor my own faith, instill hope, and respect my belief system as part of my self-care.
2) I have established helping and trusting relationships.
3) I am able to evaluate my thoughts openly and honestly no matter what my feelings are because I embrace every aspect of who I am.
4) It is important for me to create a healing environment around me that recognizes the connection between my body, mind, and spirit.
5) I appreciate myself as a whole person and seek to take care of all my needs and concerns. (Watson, 2008b).

These same five Caritas processes were retained in the caring of manager instrument as well.

It was interesting that both the caring for self instrument and the instrument we used to measure the caring of the unit manager each retained one unique item from the 10 Caritas processes that was not retained by the other. The instrument we eventually used to assess caring of the unit manager retained the Caritas item, "When my manager teaches me something new, she/he teaches me in a way that I can understand." In contrast, the caring for self instrument retained the item "I accept and support my own current beliefs in a higher power which allows for me to heal." This suggests that some Caritas processes may apply to all types of caring relationships, but also that each caring relationship may have unique aspects.

Appendix V

Factor Loading for Items in the Healing Compassions Assessment (HCA) for Use in Western Scotland

In the table below, the questions are numbered beginning with two because the first question inquired whether the survey taker was the patient or a family member. (See Table V.1.)

Table V.1 Factor loading for items in the Healing Compassions Assessment (HCA) for use in Western Scotland.

HCA #	Healing Compassion Assessment (HCA) Statement, Survey 123	Final factor loading (*n* = 504)
2	My caregivers treat me with kindness and compassion.	.764
3	My caregivers work together to meet my personal needs/ requests.	.785
4	My requests for information are treated and responded to with respect by my caregivers.	.612
5	When my caregivers talk to me about my care, condition, or treatment, they talk in a way that I can understand.	.565
6	(Caritas 5) My caregivers encourage me to practice my own individual spiritual beliefs while in their care. (Do not answer this if you feel spiritual beliefs are too personal.)	Did not load
7	My caregivers are knowledgeable.	.628
8	My caregivers are skillful.	.714
9	My caregivers have established a helping and trusting relationship with me during my time here.	.713
10, 11, 12	(Caritas 8) The environment around me helps me feel better physically; mentally; spiritually	None of these items loaded
13	My caregivers value my feelings (whatever they are) so I can talk openly and honestly about what I am thinking.	.642
Omit	Miracles	Omitted

Appendix W

Factor Loadings of the Caring Professional Scale for Use in Western Scotland

In Table W.1, the questions are numbered beginning with two because the first question inquired whether the survey taker was the patient or a family member.

Table W.1 Factor loadings of the CPS by rank.

Item	"Did you feel that the member of staff who just looked after you was:	Factor loading
2	Comforting	.933
3	Positive	.921
4	Informative	.886
5	Clinically competent	.807
6	Understanding	.886
7	Personal	.942
8	Caring	.969
9	Supportive	.954
10	An attentive listener	.889
11	Centered on you	.896
12	Technically skilled	.734
13	Aware of your feelings	Did not load
14	Visibly touched by your experience	Did not load
15	Able to offer you hope	.640
16	Respectful of you	Did not load

Appendix X

Factor Loadings for the Healing Compassion Survey—7Cs NHS Scotland (Staff Version)

All of the items in this survey began with the words, "With the patients you took care of, . . ." A factor loading shows how valid the item is, to measure what you say you want to measure. Higher factor loadings mean the item measures what you want it to measure. Note: The rows with bold text are the quality items, the lines with italicized text are the process of care items, and the rows with normal text are the caring items. (See Table X.1.)

Table X.1 Factor loadings for the Healing Compassion Survey—7Cs NHS Scotland (staff version).

For the patient you just took care of ...	Factor loading in rank order
Were you supportive?	.750
Were you caring?	.698
Did you demonstrate your commitment to quality?	**.696**
Were you an attentive listener?	.672
Were you informative?	.663
Did you display a 'can do' attitude at every opportunity?	**.657**
Did you make sure you had everything he/she needed in his/her environment of care?	*.654*
Were you understanding?	.605
Did you take responsibility to do the job well?	**.586**

Table X.1 (Continued)

For the patient you just took care of …	Factor loading in rank order
Were you personal (treated the patient kindly and as an individual)?	.568
Were you respectful of the patient?	**.555**
Were you technically skilled?	**.496**
Were you able to encourage the patients to make decisions about their care/treatment?	*.477*
Were you clinically competent?	*.467*
Did you work effectively with others in teams?	**.456**
Were you comforting?	.431
Were you able to ensure that his/her faith or belief needs were met?	*.414*
Has he/she consistently been part of your patient assignment?	*.376*

Appendix Y

Factor Analysis and Factor Ranking for Survey Items Related to Caring for Self and Caring of the Senior Charge Nurse

Using Eigenvalues greater than 1.0, factor analysis revealed one single construct of caring for self with factor loadings ranging from .752 to .919. KMO was .919 and the explained variance of caring for self was 74.9%. All items and factor loadings are noted in Table Y.1.

Using Eigenvalues greater than 1.0, factor analysis revealed one single construct of the caring of the senior charge nurse (SCN), with factor loadings ranging from .830 to .964. KMO was .950, and the explained variance of care of self was 87.3%. All items and factor loadings are noted in Table Y.2.

Table Y.1 Factor loading caring for self.

| | Factor |
Statements from Caring for Self Measure	1
Q88_CS4: I take time to consider my own thoughts.	.919
Q90_CS6: I take time to comfort myself when needed.	.912
Q87_CS3: I take time to treat myself as an individual when needed.	.911
Q86_CS2: I take time to care for myself.	.897
Q89_CS5: I take time to understand myself and reflect.	.869
Q85_CS1: I take time to find support for myself when needed.	.785
Q91_CS7: I take time to keep myself updated and informed when needed.	.752

Table Y.2 Factor loadings for caring of senior charge nurse.

	Factor
Statements from Caring of Senior Charge Nurse (SCN)	**1**
Q95_CSN: The SCN of my ward is personal (treats you as an individual).	.964
Q97_CSN: The SCN of my ward is understanding.	.960
Q96_CSN: The SCN of my ward is an attentive listener.	.955
Q94_CSN: The SCN of my ward is caring.	.951
Q93_CSN: The SCN of my ward is supportive.	.946
Q98_CSN: The SCN of my ward is comforting, when required.	.928
Q99_CSN: The SCN of my ward is informative about things related to my ward and job.	.830

Appendix Z

Demographics, Particularly Ward, as Predictors of Job Satisfaction

Table Z.1 Ward demographics as predictors of job satisfaction.

Demographic	Explained Variance of Job Satisfaction as measured by HES	p-Value (statistically significant predictor is <.001)	Grouping positively impacting job satisfaction
Ward	19.1%	< .001	Working on SDU or Nat'l Service Dept. Also, to a lesser degree, on outpatient
Education	0.9%	.303	Higher education
Gender	0.7%	.133	Being female
Marital	0.2%	.449	Being partnered (married or domestically partnered)
Hours worked	1.1%	.156	Working 36 to 40 hours or fewer than 40 hours per week (in contract to more than 40)
Age	2.1%	.073	Fewer than 30 years of age or more than 50
Unit years	3.9%	.003	Fewer than 5 years on same unit
Hospital years	0.9%	.359	Fewer than 5 years in same hospital
Professional years	1.5%	.170	Fewer than 5 years in the profession of nursing
Schedule	3.5%	.004	Working the day shift (versus 1900–0700)

Appendix AA

Demographic as Predictors of clarity

Table AA.1 Demographics as predictors of clarity.

Demographic	Explained Variance of Clarity	p-Value (statistically significant predictor is <.001)	Grouping positively impacting clarity
Ward	13.0%	<.001	Working on SDU or Nat'l Service Dept. Also, to a lesser degree, on 3 East, CCU, Outpatient, CDU or 3 West
Education	0.6%	.446	Higher education
Gender	1.2%	.041	Being female
Marital	0.0%	.823	NA. No relationship at all
Hours worked	0.4%	.461	Working 36 to 40 hr per week or fewer than 40 hr per week (in contract to more than 40)
Age	1.4%	.189	Age 40 years or more, especially if over 50
Unit years	1.3%	.198	Fewer than 5 years or more than 15 years
Hospital years	0.6%	.535	Fewer than 5 years in same hospital and 11–15 years
Professional years	1.9%	.087	Fewer than 5 years in the profession of nursing or more than 20 years
Schedule	0.3%	.087	Working the day shift (versus 1900–0700)

Appendix BB

Correlates of Operations of CBAS
with Items from the Healing Compassion
Survey – 7 Cs NHS Scotland (Staff Version)

In Tables BB.1 and BB.2, each "correlation" shows the relationship between the Operations of CBAS item in the far left column to the patient care behaviors in the columns on the right side of each table. Only correlations of .190 or above are significant, and in these tables, significant correlations are highlighted in gray. For the first Operations of CBAS item in Table BB.1, you can see that people who say they are "able to contact CBAS facilitators easily" are also people who report themselves to be "supportive" and "an attentive listener." All three of these things, in essence, increase together, 19.9% of the time.

It is important to remember here that correlation is not causation. The reason it is helpful to look at correlations such as these is mainly so the relationships identified can be studied further in predictive analytics. Is CBAS a predictor of quality? Deeper research is needed, and this initial research gives us a head start on any further studies we may undertake.

Table BB.1 Relationship between operations of CBAS items and aspects of caring (part one).

Item wording from Operations of CBAS questionnaire	Statistics	For the patient you just took care of, were you....							
		Caring and Quality—Caring total	Comforting	Informative	Understanding	Personal	Caring	Supportive	An attentive listener
I am able to contact CBAS facilitators easily.	Correlation	.208	.084	.178	.108	.180	.183	.194	.199
	Sig.	.029	.381	.062	.261	.061	.055	.042	.037
	N	110	110	110	110	110	110	110	110
I understand how the Person-centered Care Quality Instrument (PCQI) is used to improve patient care.	Correlation	.316	.164	.196	.314	.314	.297	.219	.297
	Sig. (2-tailed)	.001	.091	.043	.001	.001	.002	.023	.002
	N	108	108	108	108	108	108	108	108
I remember using the PCQI well in my CBAS training.	Correlation	.324	.155	.290	.334	.198	.258	.302	.230
	Sig.	.001	.108	.002	.000	.040	.007	.002	.016
	N	108	108	108	108	108	108	108	108
It was clear in our CBAS training how the PCQI could support our team to create action plans for change.	Correlation	.299	.093	.251	.291	.189	.231	.337	.246
	Sig.	.002	.342	.009	.003	.052	.017	.000	.011
	N	106	106	106	106	106	106	106	106
I believe CBAS is important for improving the care at this hospital.	Correlation	.221	.063	.055	.307	.221	.288	.226	.192
	Sig.	.022	.519	.573	.001	.022	.003	.019	.048
	N	107	107	107	107	107	107	107	107

(Continued)

Table BB.1 (Continued)

Item wording from Operations of CBAS questionnaire	Statistics	For the patient you just took care of, were you….							
		Caring and Quality – Caring total	Comforting	Informative	Understanding	Personal	Caring	Supportive	An attentive listener
Coworkers I work with believe CBAS is important for improving the care at this hospital.	Correlation	.143	.049	-.019	.198	.165	.200	.133	.161
	Sig.	.140	.616	.848	.040	.087	.037	.170	.097
	N	108	108	108	108	108	108	108	108
I feel there is adequate time for me to do what I feel I need to do to help implement our action plan following the CBAS training on my ward/department.	Correlation	.199	.206	.139	.144	.131	.148	.131	.171
	Sig.	.039	.033	.153	.138	.177	.127	.176	.076
	N	108	108	108	108	108	108	108	108
There are enough resources for me to feel we will be able to implement CBAS action plan on my ward.	Correlation	.153	.195	.025	.131	.144	.131	.078	.158
	Sig.	.114	.043	.799	.177	.136	.176	.422	.103
	N	108	108	108	108	108	108	108	108

The senior charge nurse I work with inspires me to apply what I learned during the CBAS training within patient care.	Correlation	.244	.061	.194	.154	.270	.134	.216	.281
	Sig	.010	.526	.042	.106	.004	.162	.023	.003
	N	111	111	111	111	111	111	111	111
The senior charge nurse is part of the support I feel is helpful to successfully implement the action plans we developed in the CBAS training.	Correlation	.222	.114	.143	.179	.232	.111	.190	.226
	Sig.	.020	.240	.137	.063	.015	.251	.048	.018
	N	109	109	109	109	109	109	109	109

Table BB.2 Relationship between operations of CBAS items and aspects of caring (part two).

Item wording from Operations of CBAS questionnaire	Statistics	Caring and Quality–Quality total	Respectful	Technically skilled	Take responsibility to do the job well	Demonstrate your commitment to quality	Work effectively with others in teams	Display a "can do" attitude at every opportunity
		For the patient you just took care of, were you						
I know who the CBAS facilitators are for my organization.	Correlation	.219	.081	.282	.095	.184	.117	.179
	Sig.	.022	.401	.003	.321	.054	.224	.061
	N	109	109	110	110	110	110	110
I am able to contact CBAS facilitators easily.	Correlation	.278	.144	.332	.148	.245	.147	.203
	Sig.	.004	.138	.000	.125	.010	.127	.034
	N	108	108	109	109	109	109	109
The CBAS facilitators are always helpful.	Correlation	.243	.145	.321	.119	.224	.104	.158
	Sig.	.011	.132	.001	.216	.019	.281	.099
	N	109	109	110	110	110	110	110
The CBAS facilitators I work with are expert in helping me understand what behaviors I can change to provide the best patient care possible.	Correlation	.201	.152	.145	.096	.146	.147	.225
	Sig.	.040	.122	.138	.326	.137	.132	.020
	N	105	105	106	106	106	106	106

I understand how the Person-centered Care Quality Instrument (PCQI) is used to improve patient care.	Correlation	.322	.303	.135	.256	.268	.275	.276
	Sig.	.001	.002	.167	.008	.005	.004	.004
	N	106	106	107	107	107	107	107
I remember using the PCQI well in my CBAS training.	Correlation	.310	.312	.143	.215	.191	.206	.387
	Sig.	.001	.001	.141	.026	.048	.033	.000
	N	106	106	107	107	107	107	107
It was clear in our CBAS training how the PCQI could support our team to create action plans for change.	Correlation	.329	.286	.277	.213	.182	.181	.368
	Sig.	.001	.003	.004	.029	.064	.064	.000
	N	104	104	105	105	105	105	105
I believe CBAS is important for improving the care at this hospital.	Correlation	.202	.257	.064	.187	.126	.143	.220
	Sig.	.039	.008	.518	.055	.198	.143	.023
	N	105	105	106	106	106	106	106
I feel there is adequate time for me to do what I feel I need to do to help implement our action plan following the CBAS training on my ward/department.	Correlation	.210	.157	.149	.065	.131	.182	.255
	Sig.	.031	.108	.126	.508	.179	.061	.008
	N	106	106	107	107	107	107	107

(Continued)

Table BB.2 (Continued)

Item wording from Operations of CBAS questionnaire	Statistics	Caring and Quality– Quality total	Respectful	Technically skilled	Take responsibility to do the job well	Demonstrate your commitment to quality	Work effectively with others in teams	Display a "can do" attitude at every opportunity
There are enough resources for me to feel we will be able to implement CBAS action plan on my ward.	Correlation	.198	.115	.159	.064	.169	.141	.227
	Sig.	.042	.242	.103	.511	.083	.148	.019
	N	106	106	107	107	107	107	107
The senior charge nurse I work with inspires me to apply what I learned during the CBAS training within patient care.	Correlation	.288	.090	.268	.274	.308	.044	.293
	Sig.	.002	.355	.005	.004	.001	.644	.002
	N	109	109	110	110	110	110	110
The senior charge nurse is part of the support I feel is helpful to successfully implement the action plans we developed in the CBAS training.	Correlation	.227	.122	.233	.210	.221	.015	.231
	Sig.	.019	.212	.015	.029	.022	.881	.016
	N	107	107	108	108	108	108	108

For the patient you just took care of, were you . . . .

References

Agostinho, J. R., Goncalves, I., Rigueira, J., Aguiar-Ricardo, I., Nunes-Ferreira, A., Santos, R., . . . Investigators, R. I.-H. (2019). Protocol-based follow-up program for heart failure patients: Impact on prognosis and quality of life. *Revista Portuguesa De Cardiologia*, 38(11), 755–764. doi:10.1016/j.repc.2019.03.006

Anandarajah, A. P., Quill, T. E., & Privitera, M. R. (2018). Adopting the quadruple aim: The University of Rochester Medical Center experience moving from physician burnout to physician resilience. *American Journal of Medicine*, 131(8), 979–986. doi:10.1016/j.amjmed.2018.04.034

Arain, M., Campbell, M. J., Cooper, C. L., & Lancaster, G. A. (2010). What is a pilot or feasibility study? A review of current practice and editorial policy. *BMC Medical Research Methodology*, 10(1), 67. doi:10.1186/1471-2288-10-67

Backonja, M., Dahl, J., Gordon, D., Rudin, N., Seghal, N., & Gravel-Sullivan, A. (2010). Pain management curriculum for Resident Physicians Module© University of Wisconsin School of Medicine and Public Health. Retrieved July 1, 2018, from http://projects.hsl.wisc.edu/GME/PainManagement/index.html

Bandura, A. (1977). Self-efficacy: Toward a unifying theory of behavioral change. *Psychological Review*, 84(2), 191. doi:10.1037/0033-295X.84.2.191

Batcheller, J., Zimmermann, D., Pappas, S., & Adams, J. M. (2017). Nursing's leadership role in addressing the quadruple aim. *Nurse Leader*, 15(3), 203–206. doi:10.1016/j.mnl.2017.02.007

Bates, D.W., Saria, S., Ohno-Machado, L., Shah, A., & Escobar, G. (2014). Big data in health care: Using analytics to identify and manage high-risk and high-cost patients. *Health Affairs*, 33(7), 1123–1131. doi:10.1377/hlthaff.2014.0041

Bayet, S., Bushnell, M. C., & Schweinhardt, P. (2014). Emotional faces alter pain perception. *European Journal of Pain*, 18(5), 712–720.

Using Predictive Analytics to Improve Healthcare Outcomes, First Edition.
Edited by John W. Nelson, Jayne Felgen, and Mary Ann Hozak.
© 2021 John Wiley & Sons, Inc. Published 2021 by John Wiley & Sons, Inc.

Bejan, A. (2019). *Freedom and evolution: Hierarchy in nature, society and science.* New York, NY: Springer Nature.

Bejan, A., & Zane, P. (2012). *Design in nature.* New York, NY: Doubleday.

Bernard, S., Chelminski, P., Ives, T., & Ranapurwala, S. (2018). Management of pain in the United States—A brief history and implications for the opioid epidemic. *Health Services Insights,* 11, 1–6. doi:10.1177/1178632918819440

Berry, D. M., Kaylor, M. B., Church, J., Campbell, K., McMillin, T., & Wamsley, R. (2013). Caritas and job environment: A replication of Persky et al. *Contemporary Nurse,* 43(2), 237–243. doi:10.5172/conu.2013.43.2.237

Berry, P. H., Chapman, C. R., Covington, E. C., Dahl, J. L., Katz, J. A., Miaskowski, C., & McLean, M. J. (2001). *Pain: Current understanding of assessment, management, and treatments* (pp. b44). VA, USA: National Pharmaceutical Council and the Joint Commission for the Accreditation of Healthcare Organizations.

Berwick, D. M., Nolan, T. W., & Whittington, J. (2008). The triple aim: Care, health, and cost. *Health Affairs,* 27(3), 759–769. doi:10.1377/hlthaff.27.3.759

Blash, L., Chan, K., & Chapman, S. (2015). *The peer provider workforce in behavioral Health: A landscape analysis.* San Francisco, CA: UCSF Health Workforce Research Center on Long-Term Care. Retrieved from https://healthworkforce.ucsf.edu/sites/healthworkforce.ucsf.edu/files/Report-Peer_Provider_Workforce_in_Behavioral_Health-A_Landscape_Analysis.pdf

Bodenheimer, T., & Sinsky, C. (2014). From triple to quadruple aim: Care of the patient requires care of the provider. *Annals of Family Medicine,* 12(6), 573–576. doi:10.1370/afm.1713

Bogetz, J. F., & Friebert, S. (2017). Defining success in pediatric palliative care while tackling the quadruple aim. *Journal of Palliative Medicine,* 20(2), 116–119. doi:10.1089/jpm.2016.0389

Bolima, D. C. (2015). *The relationship between caring leadership, nursing job satisfaction, and turnover intentions.* Grand Canyon University. Retrieved from https://search.proquest.com/openview/4485f8aa99058db040a0bf7c1d4d313c/1?pq-origsite=gscholar&cbl=18750&diss=y

Boller, J. (2017). Nurse educators: Leading health care to the quadruple aim sweet spot. *Journal of Nursing Education,* 56(12), 707–708. doi:10.3928/01484834-20171120-01

Bosserman, L. D. (2016). Pathways, processes, team work: Paving the way for value-based care with the quadruple aim. *Journal of Community and Supportive Oncology,* 14(7), 287–290. doi:10.12788/jcso.0278

Brown-Johnson, C. G., Chan, G. K., Winget, M., Shaw, J. G., Patton, K., Hussain, R., . . . Mahoney, M. (2019). Primary Care 2.0: Design of a transformational team-based practice model to meet the quadruple aim. *American Journal of Medical Quality,* 34(4), 339–347. doi:10.1177/1062860618802365

Cassel, J. B., Kerr, K., Pantilat, S., & Smith, T. J. (2010). Palliative care consultation and hospital length of stay. *Journal of Palliative Medicine*, 13(6), 761–767. doi:10.1089/jpm.2009.0379

CDC. (2017). *Wide-ranging online data for epidemiologic research (WONDER)*. Atlanta, GA: CDC, National Center for Health Statistics. Retrieved from http://wonder.cdc.gov

Centers for Disease Control and Prevention. (2018). *2018 annual surveillance report of drug-related risks and outcomes—United States*. Surveillance Special Report 2.

Centers for Disease Control and Prevention. (n.d.). *Module 5: Assessing and addressing opioid use disorder (OUD)*. Retrieved from https://www.cdc.gov/drugoverdose/training/oud/accessible/index.html

Centers for Medicare and Medicaid Services. (2018). *CAHPS® hospital survey quality assurance guide. (Version 13.0)*. Retrieved from https://www.hcahpsonline.org/globalassets/hcahps/quality-assurance/2018_qag_v13.0.pdf

Christie, C., Baker, C., Cooper, R., Kennedy, C.P.J., Madras, B., Bondi, P., & US Government Printing Office. (2017). *The President's Commission on combating drug addiction and the opioid crisis. 1*. Retrieved from https://www.whitehouse.gov/sites/whitehouse.gov/files/images/Final_Report_Draft_11-1-2017.pdf

Ciemins, E. L., Blum, L., Nunley, M., Lasher, A., & Newman, J. M. (2007). The economic and clinical impact of an inpatient palliative care consultation service: A multifaceted approach. *Journal of Palliative Medicine*, 10(6), 1347–1355. doi:10.1089/jpm.2007.0065

Ciesielski, T., Iyengar, R., Bothra, A., Tomala, D., Cislo, G., & Gage, B. (2016). A tool to assess risk of de novo opioid abuse or dependence. *The American Journal of Medicine*, 129(7), 699–705. e694. doi:10.1016/j.amjmed.2016.02.014

Coates, C. (2019). Caring efficacy scale. In K. L. Sitzman & J. Watson (Eds.), *Assessing and measuring caring in nursing and health sciences* (3rd ed., pp. 165–171). New York: Springer.

Coates, C. J. (1997). The caring efficacy scale: Nurses' self-reports of caring in practice settings. *Advanced Practice Nursing Quarterly*, 3(1), 53–59.

Cochran, B. N., Flentje, A., Heck, N. C., Van Den Bos, J., Perlman, D., Torres, J., . . . Carter, J. (2014). Factors predicting development of opioid use disorders among individuals who receive an initial opioid prescription: mathematical modeling using a database of commercially-insured individuals. *Drug and Alcohol Dependence*, 138, 202–208. doi:10.1016/j.drugalcdep.2014.02.701

Cowan, J. D. (2004). Hospital charges for a community in patient palliative care program. *American Journal of Hospice and Palliative Care*, 21(3), 177–190. doi:10.1177/104990910402100306

Creative Health Care Management. (2017). *A quick guide to relationship-based care.* Minneapolis, MN: Creative Health Care Management.

Davenport, T. H. (2006). Competing on analytics. *Harvard Business Review*, 84(1), 98. Retrieved from https://hbr.org/2006/01/competing-on-analytics

Davis, J., Kutash, M., & Whyte, J. (2017). A comparative study of patient sitters with video monitoring versus in-room sitters. *Journal of Nursing Education*, 7(3), 137–142. doi:10.5430/jnep.v7n3p137

Dempsey, C. (2014). *Quality improvement and evidence-based practice—making the connection: Reducing suffering with compassionate connected care* [Presentation slides]. Retrieved from https://www.inova.org/upload/docs/For-Nurses/Nursing%20Research/Dempsey_Inova_Dec2014.pdf

DiNapoli, P. P., Nelson, J., Turkel, M., & Watson, J. (2010). Measuring the Caritas Processes: Caring Factor Survey. *International Journal of Human Caring*, (3), 15–20. doi:10.20467/1091-5710.14.3.15

Donroe, J. H., Holt, S. R., & Tetrault, J. M. (2016). Caring for patients with opioid use disorder in the hospital. *CMAJ: Canadian Medical Association Journal*, 188(17–18), 1232.

Duclay, E., Hardouin, J. B., Sebille, V., Anthoine, E., & Moret, L. (2015). Exploring the impact of staff absenteeism on patient satisfaction using routine databases in a university hospital. *Journal of Nursing Management*, 23(7), 833–841. doi:10.1111/jonm.12219

Duffy, J. R., Brewer, B. B., & Weaver, M. T. (2014). Revision and psychometric properties of the caring assessment tool. *Clinical Nursing Research*, 23(1), 80–93. doi:10.1177/1054773810369827

Duncan, E. A. S., Colver, K., Stephenson, J., & Abhyanakar, P. (2017). *Scottish person centredness improvement collaboration (SCOPIC).* Stirling: The nursing, midwifery and allied health professions research unit (NMAHP RU). Retrieved from http://www.nmahp-ru.ac.uk/media/microsites/nmahp-ru/documents/1718038-NMAHP-RU-Book-2-WEB-Singles-FINAL.pdf

Dunlap, L. J., Zarkin, G. A., Orme, S., Meinhofer, A., Kelly, S. M., O'Grady, K. E., . . . Schwartz, R. P. (2018). Re-engineering methadone-cost-effectiveness analysis of a patient-centered approach to methadone treatment. *Journal of Substance Abuse Treatment*, 94, 81–90. doi:10.1016/j.jsat.2018.07.014

Edmundson, E. (2012). The quality caring nursing model: A journey to selection and implementation. *Journal of Pediatric Nursing-Nursing Care of Children & Families*, 27(4), 411–415. doi:10.1016/j.pedn.2011.09.007

Eisler, R., & Potter, T. (2014). *Transforming interprofessional partnerships: A new framework for nursing and partnership-based health care.* Indianapolis, IN: Sigma Theta Tau International.

Falen, T., Alexander, J., Curtis, D., & UnRuh, L. (2013). Developing a hospital-specific electronic inpatient fall surveillance program: Phase 1. *Health Care Manager*, 32(4), 359–369. doi:10.1097/HCM.0b13e3182a9d6ec

Fallin-Bennett, A., Elswick, A., & Ashford, K. (2020). Peer support specialists and perinatal opioid use disorder: Someone that's been there, lived it, seen it. *Addictive behaviors*, 102. doi:10.1016/j.addbeh.2019.106204

Felgen, J. (2007). *I2E2: Leading lasting change*. Minneapolis, MN: Creative Health Care Management.

Felgen, J., & Koloroutis, M. (2007). The five conditions for engaging in change: The 5 Cs. In M. Koloroutis, J. Felgen, C. Person, & S. Wessel (Eds.), *Relationship-based care: Visions, strategies, tools and exemplars for transforming practice* (pp. 480–482). Minneapolis, MN: Creative Health Care Management.

Felgen, J., & Nelson, J. W. (2016). *Parallel and factor analysis for delimiting items in a three-dimensional construct of clarity of nurses*. (Publication no. http://hdl.handle.net/10755/621134). Sigma Theta Tau International (STTI).

Ganann, R., Weeres, A., Lam, A., Chung, H., & Valaitis, R. (2019). Optimization of home care nurses in Canada: A scoping review. *Health & Social Care in the Community*, 27(5), E604–E621. doi:10.1111/hsc.12797

Gaudine, A., & Gregory, C. (2010). The accuracy of nurses' estimates of their absenteeism. *Journal of Nursing Management*, 18(5), 599–605. doi:10.1111/j.1365-2834.2010.01107.x

Goodnow, M. (1929). *Outlines of nursing history* (4th ed.). Philadelphia, PA: Saunders.

Gorman, E., Yu, S. C., & Alamgir, H. (2010). When healthcare workers get sick: Exploring sickness absenteeism in British Columbia, Canada. *Work: A Journal of Prevention Assessment & Rehabilitation*, 35(2), 117–123. doi:10.3233/wor-2010-0963

Gorski, T., & Miller, M. (1986). *Staying sober: A guide for relapse prevention*. Independence, MO: Herald Publishing House.

Gözüm, S., Nelson, J., Yildirium, N., & Kavla, I. (2021). Translation and psychometric testing of the Healthcare Environment Survey in Turkey. *Florence Nightingale Journal of Nursing*, 29(1), 103–112. doi:10.5152/FNJN.2021.20014.

Griffeth, R. W., Hom, P. W., & Gaertner, S. (2000). A meta-analysis of antecedents and correlates of employee turnover: Update, moderator tests, and research implications for the next millennium. *Journal of Management*, 26(3), 463–488. doi:10.1177/014920630002600304

Guanci, G., & Medeiros, M. (2018). *Shared governance that works*. Minneapolis, MN: Creative Health Care Management.

Haddad, Y. K., Bergen, G., & Florence, C. S. (2019). Estimating the economic burden related to older adult falls by state. *Journal of Public Health Management and Practice*, 25(2), E17–E24. doi:10.1097/phh.0000000000000816

Hahn, L., Belisle, M., Nguyen, S., Alvarez, K. S., & Das, S. (2019). Effect of pharmacist clinic visits on 30-day heart failure readmission rates at a county hospital. *Hospital Pharmacy*, 54(6), 358–364. doi:10.1177/0018578718797263

Halter, M., Boiko, O., Pelone, F., Beighton, C., Harris, R., Gale, J., . . . Drennan, V. (2017). The determinants and consequences of adult nursing staff turnover: A systematic review of systematic reviews. *BMC Health Services Research*, 17. doi:10.1186/s12913-017-2707-0

Hamilton, J. (1989). Comfort and the hospitalized chronically ill. *Journal of Gerontological Nursing*, 15(4), 28–33. doi:10.3928/0098-9134-19890401-08

Hani, J. (2017). *The neuroscience of behavior change: Helping patients change behavior by understanding the brain*. Retrieved from https://healthtransformer.co/the-neuroscience-of-behavior-change-bcb567fa83c1

Hansen, G. R., & Streltzer, J. (2005). The psychology of pain. *Emergency Medicine Clinics of North America*, 23(2), 339–348.

Hanson, R. (2018). *Resilient: How to grow an unshakeable core of calm, strength, and happiness*. New York: Harmony Books.

Havens, D. S., Gittell, J. H., & Vasey, J. (2018). Impact of relational coordination on nurse job satisfaction, work engagement and burnout: Achieving the quadruple aim. *Journal of Nursing Administration*, 48(3), 132–140. doi:10.1097/nna.0000000000000587

Hazelden Betty Ford Foundation. (2020). *Connection recovery coaching*. Hazelden Betty Ford Foundation. Retrieved from https://www.hazeldenbettyford.org/recovery/preventing-relapse/connection-recovery-coaching

Health Education England. (2020). *Person-centered care*. Retrieved from https://www.hee.nhs.uk/our-work/person-centred-care

Hertzog, M. A. (2008). Considerations in determining sample size for pilot studies. *Research in Nursing & Health*, 31(2), 180–191. doi:10.1002/nur.20247

Hooper, D., Coughlan, J., & Mullen, M. R. (2008). Structual equation modeling: Guidelines for determining model fit. *Electronic Journal of Business Research Methods*, 6(1), 53-60. Retrieved from https://academic-publishing.org/index.php/ejbrm/article/view/1224

Hozak, M. A., & Brennan, M. (2012). Caring at the core: Maximizing the likelihood that a caring moment will occur. In J. W. Nelson & J. Watson (Eds.), *Measuring caring: A compilation of international research on caritas as healing* (pp. 195–223). New York, NY: Springer.

Hu, L.-t., & Bentler, P. M. (1999). Cutoff criteria for fit indexes in covariance structure analysis: Conventional criteria versus new alternatives. *Structural Equation Modeling*, 6(1), 1-55. doi:10.1080/10705519909540118

Ibrahim, M. A., Aziz, A. A., Suhaili, N. A., Daud, A. Z., Naing, L., & Rahman, H. A. (2019). A study into psychosocial work stressors and health care productivity. *International Journal of Occupational and Environmental Medicine*, 10(4), 185–193. doi:10.15171/ijoem.2019.1610

Institute of Medicine. (2001). *The future of nursing: Leading change, advancing health*. Washington, DC: The National Academies Press.

Institute of Medicine. (2011). *Report from the committee on advancing pain research, care, and education: Relieving pain in America, a blueprint for transforming prevention, care, education and research.* The National Academies Press. Retrieved from http://books.nap.edu/openbook.php?record_id=13172&page=1

Itzhaki, M., Treacy, M., Phaladze, N., Rumeu, C., Vernon, R. A., Marshall, B., . . . Nelson, J. (2015). Caring International Research Collaborative: A five-country partnership to measure perception of nursing staffs' compassion fatigue, burnout, and caring for self. *Interdisciplinary Journal of Partnership Studies*, 2(8), 1–20. doi:10.24926/ijps.v2i1.104

Itzoe, M., & Guarnieri, M. (2017). New developments in managing opioid addiction: Impact of a subdermal buprenorphine implant. *Drug Design, Development and Therapy*, 11, 1429–1437. doi:10.2147/DDDT.S109331

Jack, H. E., Oller, D., Kelly, J., Magidson, J. F., & Wakeman, S. E. (2018). Addressing substance use disorder in primary care: The role, integration, and impact of recovery coaches. *Substance Abuse*, 39(3), 307–314. doi:10.1080/08897077.2017.1389802

Jacobs, B., McGovern, J., Heinmiller, J., & Drenkard, K. (2018). Engaging employees in well-being: Moving from the triple aim to the quadruple aim. *Nursing Administration Quarterly*, 42(3), 231–245. doi:10.1097/naq.0000000000000303

Joe, G. W., Knight, D. K., Becan, J. E., & Flynn, P. M. (2014). Recovery among adolescents: Models for post-treatment gains in drug abuse treatments. *Journal of Substance Abuse Treatment*, 46(3), 362–373. doi:10.1016/j.jsat.2013.10.007

Johanson, G. A., & Brooks, G. P. (2010). Initial scale development: Sample size for pilot studies. *Educational and Psychological Measurement*, 70(3), 394–400. doi:10.1177/0013164409355692

Johnson, J. (2011). Creation of the caring factor survey - care provider version. In J. W. Nelson & J. Watson (Eds.), *Measuring caring: International research on caritas as healing.* New York: Springer.

Kahneman, D. (2011). *Thinking, fast and slow.* New York: Farrar, Straus and Giroux.

Kendi, I. (2017). *Stamped from the beginning: The definitive history of racist ideas in American.* New York, NY: Bold Type Books.

Kerfoot, K. (2006). Reliability between nurse managers: The key to the high-reliability organization. *Nursing Economics*, 24(5), 274–275. Retrieved from www.annanurse.org/download/reference/update/Volume37_Issue1/pages25-26.pdf

Kistin, C., & Silverstein, M. (2015). Pilot studies: A critical but potentially misused component of interventional research. *The Journal of the American Medical Association*, 314(15), 1561–1562. doi:10.1001/jama.2015.10962

Koch, E. (1999). Alice Magaw and the great secret of open drop anesthesia. *Aana Journal*, 67(1), 33–34. Retrieved from https://www.aana.com/docs/default-source/exec-unit-aana-com-web-documents-(all)/archives-library/imaging-in-time---alice-magaw.pdf?sfvrsn=f09c45b1_4

Kolcaba, K. (2003). *Comfort theory and practice: A vision for holistic health care and research*. New York, NY: Springer Publishing Company.

Kolcaba, K. Y. (1991). A taxonomic structure for the concept comfort. *Image: The Journal of Nursing Scholarship, 23*(4), 237–240. doi:10.1111/j.1547-5069.1991.tb00678.x

Koloroutis, M. (Ed.) (2004). *Relationship-based care: A model for transforming practice*. Minneapolis, MN: Creative Health Care Management.

Koloroutis, M., & Abelson, D. (Eds.) (2017). *Advancing relationship-based cultures*. Minneapolis, MN: Creative Health Care Management.

Koloroutis, M., Felgen, J., Person, C., & Wessel, S. (Eds.) (2007). *The relationship-based care field guide: Visions, strategies, tools and exemplars for transforming practice*. Minneapolis, MN: Creative Health Care Management.

Koloroutis, M., & Trout, M. (2012). *See me as a person: Creating therapeutic relationships with patients and their families*. Minneapolis, MN: Creative Health Care Management.

Kramer, M., Schmalenberg, C., Brewer, B. B., Verran, J. A., & Keller-Unger, J. (2009). Accurate assessment of clinical nurses' work environments: Response rate needed. *Research in Nursing & Health, 32*(2), 229–240. doi:10.1002/nur.20315

Kuhn, T. (1962). *The structure of scientific revolution*. Chicago, IL: Chicago Press.

Kurowski, C., Murakami, Y., Ono, T., Shors, L., Vujicie, M., & Zolfaghari, A. (2009). *The nurse labor and education markets in the English-speaking CARICOM: Issues and options for reform*. Washington, DC: World Bank.

Lancaster, G. A. (2015). Pilot and feasibility studies come of age! *Pilot and Feasibility Studies, 1*(1), 1. doi:10.1186/2055-5784-1-1

Langford, A. T. (2017). *Best practices in prescribing opioids for chronic pain*. Retrieved from www.rheumatologynetwork.com/best-practices-prescribing-opioids-chronic-pain

Lawrence, I., & Kear, M. (2011). The practice of loving kindness to self and others as perceived by nurses and patients in the cardiac interventional unit. In J. W. Nelson & J. Watson (Eds.), *Measuring caring: International research on caritas as healing* (pp. 36–39). New York, NY: Springer.

Lazenby, M. (2017). *Caring matters most: The ethical significance of nursing*. New York, NY: Oxford University Press.

Leavitt, M. A., Hain, D. J., Keller, K. B., & Newman, D. (2020). Testing the effect of a home health heart failure intervention on hospital readmissions, heart failure knowledge, self-care, and quality of life. *Journal of Gerontological Nursing, 46*(2), 32-+. doi:10.3928/00989134-20191118-01

Lee, J. Y., Jin, Y., Piao, J., & Lee, S. M. (2016). Development and evaluation of an automated fall risk assessment system. *International Journal for Quality in Health Care, 28*(2), 175–182. doi:10.1093/intqhc/mzv122

Leon, A. C., Davis, L. L., & Kraemer, H. C. (2011). The role and interpretation of pilot studies in clinical research. *Journal of Psychiatric Research*, 45(5), 626–629. doi:10.1016/j.jpsychires.2010.10.008

Lever, I., Dyball, D., Greenberg, N., & Stevelink, S. A. M. (2019). Health consequences of bullying in the healthcare workplace: A systematic review. *Journal of Advanced Nursing*, 75(12), 3195–3209. doi:10.1111/jan.13986

Lewis, M. (2004). *Moneyball: The art of winning an unfair game*. New York, NY: WW Norton & Company.

Li, Y., & Jones, C. B. (2013). A literature review of nursing turnover costs. *Journal of Nursing Management*, 21(3), 405–418. doi:10.1111/j.1365-2834.2012.01411.x

Longbrake, K. (2017). Focusing on the quadruple aim of health care. *American Nurse Today*, 12(7), 44–46. Retrieved from https://www.americannursetoday.com/wp-content/uploads/2017/06/ant7-Quadruple-626a.pdf

Mahajan, S. M., Heidenreich, P., Abbott, B., Newton, A., & Ward, D. (2018). Predictive models for identifying risk of readmission after index hospitalization for heart failure: A systematic review. *European Journal of Cardiovascular Nursing*, 17(8), 675–689. doi:10.1177/1474515118799059

Manthey, M. (1980). *The practice of primary nursing*. Minneapolis, MN: Creative Health Care Management.

Manthey, M. (2007). Responsibility, authority, and accountability. In M. Koloroutis, J. Felgen, C. Person, & S. Wessel (Eds.), *Relationship-based care: Visions, strategies, tools and exemplars for transforming practice* (pp. 486). Minneapolis, MN: Creative Health Care Management.

Martin, K. D., Cullen, J. B., & Parboteeah (2007). Deciding to bribe: A cross-level analysis of firm and home country influences on bribary activity. *Academy of Management Journal*, 50(6), 1401–1422. doi:10.5465/AMJ.2007.28179462

Martinez, A., & Allen, A. (2020). A review of nonpharmacological adjunctive treatment for postpartum women with opioid use disorder. *Addictive Behaviors*, 105, 106323–106323. doi:10.1016/j.addbeh.2020.106323

May, P., Garrido, M. M., Cassel, J. B., Kelley, A. S., Meier, D. E., Normand, C., . . . Morrison, R. S. (2017). Cost analysis of a prospective multi-site cohort study of palliative care consultation teams for adults with advanced cancer: Where do cost-savings come from? *Palliative Medicine*, 31(4), 378–386. doi:10.1177/0269216317690098

Mayr, F. B., Talisa, V. B., Balakumar, V., Chang, C.-C. H., Fine, M., & Yende, S. (2017). Proportion and cost of unplanned 30-day readmissions after sepsis compared with other medical conditions. *The Journal of the American Medical Association*, 317(5), 530–531. doi:10.1001/jama.2016.20468%J JAMA

McDonald, L. (2001). Florence Nightingale and the early origins of evidence-based nursing. *Evidence Based Nursing*, 4(3), 68. doi:10.1136/ebn.4.3.68

McInnes, S., Peters, K., Bonney, A., & Halcomb, E. (2017). A qualitative study of collaboration in general practice: Understanding the general practice nurse's role. *Journal of Clinical Nursing*, 26(13–14), 1960–1968. doi:10.1111/jocn.13598

Mee-Lee, D. (2013). *The ASAM criteria: Treatment criteria for addictive, substance-related, and co-occurring conditions*. Chevy Chase, MD: American Society of Addiction Medicine.

Mehta, V., & Langford, R. M. (2006). Acute pain management for opioid dependent patients. *Anaesthesia*, 61(3), 269–276.

Melnick, E. R., & Powsner, S. M. (2016). Empathy in the time of burnout. *Mayo Clinic Proceedings*, 91(12), 1678–1679. doi:10.1016/j.mayocp.2016.09.003

Melzack, R. (2001). Pain and the neuromatrix in the brain. *Journal of Dental Education*, 65(12), 1378–1382.

Melzack, R., & Wall, P. D. (1965). Pain mechanisms: A new theory. *Science*, 150, 971–979.

Merskey, H., & Bugduk, N. (1994). *Classification of chronic pain: Descriptions of chronic pain syndromes and definitions of pain terms* (2nd ed.). Seattle, WA: IASP Press.

Messina, B. G., & Worley, M. J. (2019). Effects of craving on opioid use are attenuated after pain coping counseling in adults with chronic pain and prescription opioid addiction. *Journal of Consulting and Clinical Psychology*, 87(10), 918–926. doi:10.1037/ccp0000399

Mizukawa, M., Moriyama, M., Yamamoto, H., Rahman, M. M., Naka, M., Kitagawa, T., . . . Kihara, Y. (2019). Nurse-led collaborative management using telemonitoring improves quality of life and prevention of rehospitalization in patients with heart failure a pilot study. *International Heart Journal*, 60(6), 1293–1302. doi:10.1536/ihj.19-313

National Institute for Health and Care Excellence. (2014). *Acute heart failure: Diagnosis and management*. Retrieved from https://www.nice.org.uk/guidance/cg187

National Institute for Health and Care Excellence. (2018). *Chronic heart failure in adults: Diagnosis and management*. Retrieved from https://www.nice.org.uk/guidance/ng106

Nelson, J. W., Valentino, L., Iacono, L., Ropollo, P., Cineas, N., & Stuart, S. (2015). Measuring workload of nurses on a neurosurgical care unit. *Journal of Neuroscience Nursing*, 47(3). doi:10.1097/JNN.0000000000000136

Nelson, J. W. (2001). *A professional nursing care model and satisfaction of the staff nurse. (Masters)*. Minneapolis, MN: University of Minnesota.

Nelson, J. W. (2013). *Job satisfaction of nurses in Jamaica. (PhD)*. Minneapolis: University of Minnesota. Retrieved from https://conservancy.umn.edu/bitstream/handle/11299/162507/Nelson_umn_0130E_14514.pdf?sequence=1

Nelson, J. W., & Cavanagh, A. M. (2017). Development of an international tool to measure nurse job satisfaction by testing the Healthcare Environment Survey beyond Jamaica and the United States to Scotland: A cross sectional study utilizing exploratory factor analysis. *International Journal of Healthcare Management*, 11(1), 1–5. doi:10.1080/20479700.2017.1312803

Nelson, J. W., DiNapoli, P., Turkel, M. C., & Watson, J. (2011). Concepts of caring as construct of Caritas hierarchy in nursing knowledge: conceptual-theoretical-empirical (CTE). In J. W. Nelson & J. Watson (Eds.), *Measuring caring: A compilation of international research on caritas as healing* (chapter 3, pp. 3–18). New York: Springer.

Nelson, J. W., & Felgen, J. (2015). *Integration of caring science research across settings and implications for practice.* Paper presented at the Sigma Theta Tau International 26th Research Congress, Puerto Rico. Retrieved from http://hdl.handle. net/10755/601544

Nelson, J.W., Gallager, T., Cummings, C., Kaya, A, Nichols, T., Thomas, T., & Vrbnjak, D. (2020). *A replication systematic literature review of nurse job satisfaction, 2016–2017.* Unpublished manuscript.

Nelson, J. W., Itzhaki, M., Ehrenfeld, M., Tinker, A., Hozak, M., & Johnson, S. (2011). Nurses' caring for self, a four-country descriptive study (England, Israel, New Zealand and the USA). In J. W. Nelson & J. Watson (Eds.), *Measuring caring: A compilation of international research on caritas as healing.* New York: Springer.

Nelson, J. W., Nichols, T., & Wahl, J. (2017). The cascading effect of civility on outcomes of clarity, job satisfaction, and caring for patients. *Interdisciplinary Journal of Partnership Studies*, 4(2). Retrieved from https://pubs.lib.umn.edu/ index.php/ijps/article/view/164

Nelson, J. W., Persky, G., Hozak, M. A., Albu, A., Hinds, P. S., & Savik, K. (2015). A multistudy validation of an instrument for nurse job satisfaction. *Virginia Henderson Global Nursing e-Repository*. Retrieved from http://hdl.handle. net/10755/583356

Nelson, J. W., Thiel, L., Hozak, M. A., & Thomas, T. (2016). Item reduction of the caring factor survey – care provider version, an instrument specified to measure Watson's 10 processes of caring. *International Journal for Human Caring*, 20(3), 123–128. doi:10.20467/1091-5710-20.3.123

Nelson, J. W., & Watson, J. (2019). Development of the caring factor survey (CFS), an instrument to measure patient's perception of caring. In K. L. Sitzman & J. Watson (Eds.), *Assessing and measuring caring in nursing and health sciences* (3rd ed., pp. 271–279). New York: Springer Publishing Company.

Nelson, J. W., Vrbnjak, D., Thomas, P., Gözüm, S., Barros, A., & Itzhaki, M. (in review). The Healthcare Environment Survey: A multicountry psychometric evaluation of nurses' job satisfaction.

Neuhauser, D. (2003). Florence Nightingale gets no respect: As a statistician that is. *Quality and Safety in Health Care*, 12(4), 317. doi:10.1136/qhc.12.4.317

Nichols, T. (2017). See me as a person therapeutic practices: Core competencies applied to pain and comfort care. In M. Koloroutis & D. Abelson (Eds.), *Advancing relationship-based cultures* (pp. 273). Minneapolis, MN: Creative Health Care Management.

Nichols, T. (2018). Comfort as a multidimensional construct for pain management. *Creative Nursing*, 24(2), 88–98. doi:10.1891/1078-4535.24.2.88

Nichols, T. (2019). The role of the doctor of nursing practice in promoting nonpharmacologic pain and comfort management. *Creative Nursing*, 25(4), e28–e35.

Nichols, T., Hozak, M. A., & Nelson, J. W. (2016). *Studying HCAHPS scores and patient falls in the context of caring science*. Paper presented at the Sigma Theta Tau International (STTI) 26th Research Congress, San Juan, Puerto Rico. Retrieved from http://hdl.handle.net/10755/601530

Nightingale, F. (1959). *Notes on nursing*. Philadelphia: Lippincott.

Norton, S. A., Hogan, L. A., Holloway, R. G., Temkin-Greener, H., Buckley, M., & Quill, T. E. (2007). Proactive palliative care in the medical intensive care unit: Effects on length of stay for selected high-risk patients. *Critical Care Medicine*, 35(6), 1530–1535. doi:10.1097/01.CCM.0000266533.06543.0C

O'Grady, N. P., Alexander, M., Burns, L. A., Dellinger, E. P., Garland, J., Heard, S. O., . . . Pearson, M. L. (2011). Guidelines for the prevention of intravascular catheter-related infections. *Clinical Infectious Diseases*, 52(9), e162–e193. Retrieved from https://www.cdc.gov/infectioncontrol/guidelines/bsi/

Olender, L., & Phifer, S. (2011). Development of the caring factor survey, caring of the manager. In J. W. Nelson & J. Watson (Eds.), *Measuring caring: International research on caritas as healing* (pp. 57–62). New York, NY: Springer.

Olmstead, T. A., Abraham, A. J., Martino, S., & Roman, P. M. (2012). Counselor training in several evidence-based psychosocial addiction treatments in private US substance abuse treatment centers. *Drug and Alcohol Dependence*, 120(1-3), 149–154. doi:10.1016/j.drugalcdep.2011.07.017

Olsen, Y. (2016). The CDC guideline on opioid prescribing: Rising to the challenge. *The Journal of the American Medical Association*, 315(15), 1577–1579.

Ossipov, M., Morimura, K., & Porreca, F. (2014). Descending pain modulation and chronification of pain. *Current Opinion in Supportive and Palliative Care*, 8(2), 143–151. doi:10.1097/SPC.0000000000000055

Pasero, C., & McCaffery, M. (2010). *Pain assessment and pharmacologic management-e-book*. New York, NY: Elsevier Health Sciences.

Persky, G., Felgen, J., & Nelson, J. W. (2011). Measuring caring in primary nursing. In J. W. Nelson & J. Watson (Eds.), *Measuring caring: International research on caritas as healing* (pp. 65–86). New York: Springer.

Persky, G., Nelson, J. W., Watson, J., & Bent, K. (2008). Creating a profile of a nurse effective in caring. *Nursing Administration Quarterly*, 32(1), 15–20.

Person, C. (2004). Patient care delivery. In M. Koloroutis (Ed.), *Relationship-based care: A model for transforming practice* (pp. 159–182). Minneapolis, MN: Creative Health Care Management.

Planetree. (2020). *Person-centered care certification*. Retrieved from https://www. planetree.org/certification/planetree-certification

Ponikowski, P., Voors, A. A., Anker, S. D., Bueno, H., Cleland, J. G., Coats, A. J., . . . van der Meer, P. (2016). 2016 ESC guidelines for the diagnosis and treatment of acute and chronic heart failure: The task force for the diagnosis and treatment of acute and chronic heart failure of the European Society of Cardiology (ESC). Developed with the special contribution of the Heart Failure Association (HFA) of the ESC. *European Journal of Heart Failure*, 18(8), 891–975. doi:10.1002/ejhf.592

Pougiales, J. (1970). The first anesthetizers at the Mayo Clinic. *Journal of the American Association of Nurse Anesthetists*, 38(3), 235–241. Retrieved from https://www.aana.com/docs/default-source/exec-unit-aana-com-web-documents-(all)/archives-library/pougiales_first_anesthetizers_mayo_clinic.pdf?sfvrsn=3cd730f9_5

Privitera, M. R. (2018). Addressing human factors in burnout and the delivery of healthcare: Quality & safety imperative of the quadruple aim. *Health*, 10(05), 629. doi:10.4236/health.2018.105049

Rai, D., Zitko, P., Jones, K., Lynch, J., & Araya, R. (2013). Country-and individual-level socioeconomic determinants of depression: multilevel cross-national comparison. *The British Journal of Psychiatry*, 202(3), 195–203. doi:10.1192/bjp.bp.112.112482

Rainville, P., Bao, Q. V. H., & Chrétien, P. (2005). Pain-related emotions modulate experimental pain perception and autonomic responses. *Pain*, 118(3), 306–318.

Rathert, C., Williams, E. S., & Linhart, H. (2018). Evidence for the quadruple aim: A systematic review of the literature on physician burnout and patient outcomes. *Medical Care*, 56(12), 976–984. doi:10.1097/mlr.0000000000000999

Reese, R. L., Clement, S. A., Syeda, S., Hawley, C. E., Gosian, J. S., Cai, S., . . . Driver, J. A. (2019). Coordinated-transitional care for veterans with heart failure and chronic lung disease. *Journal of the American Geriatrics Society*, 67(7), 1502–1507. doi:10.1111/jgs.15978

Rezapour, T., Hatami, J., Farhoudian, A., Sofuoglu, M., Noroozi, A., Daneshmand, R., . . . Ekhtiari, H. (2019). Cognitive rehabilitation for individuals with opioid use disorder: A randomized controlled trial. *Neuropsychological Rehabilitation*, 29(8), 1273–1289. doi:10.1080/09602011.2017.1391103

Sackett, D. L., Rosenberg, W. M., Gray, J. A., Haynes, R. B., & Richardson, W. S. (1996). Evidence based medicine: What it is and what it isn't. *The BMJ*, 312(7023), 71–72. doi:10.1136/bmj.312.7023.71

Salmon, M. E., Yan, J., Hewitt, H., & Guisinger, V. (2007). Manages migration: The Caribbean approach to addressing nursing services capacity. *Health Services Research*, 42(3), 1354–1372. doi:10.1111/j.1475-6773.2007.00708.x

Scott, C. K., Dennis, M. L., & Foss, M. A. (2005). Utilizing recovery management checkups to shorten the cycle of relapse, treatment reentry, and recovery. *Drug and Alcohol Dependence*, 78(3), 325–338. doi:10.1016/j.drugalcdep.2004.12.005

Scott, C. K., Grella, C. E., Nicholson, L., & Dennis, M. L. (2018). Opioid recovery initiation: Pilot test of a peer outreach and modified recovery management checkup intervention for out-of-treatment opioid users. *Journal of substance abuse treatment*, 86, 30–35. doi:10.1016/j.drugalcdep.2004.12.005

Shiraly, R., & Taghva, M. (2018). Factors associated with sustained remission among chronic opioid users. *Addiction & Health*, 10(2), 86–94. doi:10.22122/ahj.v10i2.569

Sightes, E., Watson, D.P., Ray, B., Robison, L., Childress, S., & Anderson, M. (2017, October 20). The Use of peer recovery coaches to combat barriers to opioid use disorder treatment in Indiana. Retrieved from https://fsph.iupui.edu/research-centers/centers/cheer/research.html

Sikka, R., Morath, J. M., & Leape, L. (2015). *The quadruple aim: Care, health, cost and meaning in work.* BMJ Publishing Group Ltd. doi:10.1136/bmjqs-2015-004160

Sinclair, D. C. (1955). Cutaneous sensation and the doctrine of specific energy. *Brain*, 78(4), 584–614.

Sinek, S. (2011). *Start with why: How great leaders inspire everyone to take action.* New York, NY: Portfolio Penguin.

Sinek, S. (2017). *Leaders eat last: Why some teams pull together and others don't.* New York, NY: Portfolio Penguin.

Sinek, S. (2019). *The infinite game.* New York, NY: Portfolio Penguin.

Smith, D. (2018, March 16). *Implementing patient- and family-centered care models.* Retrieved from https://consultqd.clevelandclinic.org/implementing-patient-and-family-centered-care-models/

Sofuoglu, M., DeVito, E. E., & Carroll, K. M. (2019). Pharmacological and behavioral treatment of opioid use disorder. *Psychiatric Research and Clinical Practice*, 1(1), 4–15. doi:10.1176/appi.prcp.20180006

Stam, E. (2013). Knowledge and entrepreneurial employees: A country-level analysis. *Small Business Economics*, 41(4), 887–898. doi:10.1007/s11187-013-9511-y

Stroh, D. P. (2015). *Systems thinking for social change: A practical guide to solving complex problems, avoiding unintended consequences, and achieving lasting results.* White River Junction, VT: Chelsea Green Publishing.

Substance Abuse and Mental Health Services Administration (SAMHSA). (2020). *TIP 63: Medications for opioid use disorder*. Retrieved from https://store.samhsa. gov/product/TIP-63-Medications-for-Opioid-Use-Disorder-Full-Document/ PEP20-02-01-006

Sullivan, M. J. L., Bishop, S. R., & Pivik, J. (1995). The pain catastrophizing scale: Development and validation. *Psychological Assessment*, 7, 524–532.

Swanson, K. (2008). Caring professional scale. In J. Watson (Ed.), *Assessing measuring caring in nursing health science* (pp. 203–206). New York, NY: Springer.

Swanson, K. M. (1999). Effects of caring, measurement, and time on miscarriage impact and women's well-being. *Nursing Research*, 48(6), 288–298. doi:10.1097/00006199-199911000-00004

Sweet, W. H. (1959). Handbook of physiology, section 1. *Neurophysiology*, 1, 459.

Tabachnick, G. H., & Fidell, L. S. (2007). *Using multivariate statistics* (4th ed.). Boston, MA: Allyn & Bacon.

Taylor, R. (2008). *The intentional relationship: Occupational therapy and use of self*. Philadelphia, PA: F.A. Davis Company.

The Scottish Government. (2010). *The healthcare quality strategy for NHS Scotland*. Edinburgh: Scottish Government. Retrieved from http://www.gov.scot/resource/ doc/311667/0098354.pdf

Trist, E., & Emery, F. (2005). Socio-technical systems theory. In *Organizational behavior 2: Essential theories of process and structure* (Vol. 169). New York, NY: M. E. Sharpe in White Plains.

Trist, E. L., & Bamforth, K. W. (1951). Some social and psychological consequences of the longwall method of coal-getting. *Human Relations*, 4(8), 3–38. doi:10.1177/001872675100400101

Upstream. (2020). *Enabling cross-institutional collaboration to improve cancer care in central Texas*. Retrieved from https://www.upstreamthinking.com/health-projects/ creating-a-model-for-patient-centered-care

Valentine, C. M. (2018). Tackling the quadruple aim helping cardiovascular professionals find work-life balance. *Journal of the American College of Cardiology*, 71(15), 1707–1709. doi:10.1016/j.jacc.2018.03.014

Vermuri, A. W., & Costanza, R. (2006). The role of human, social, built, and natural capital in explaining life satisfaction at the country level: Toward a national well-being index (NWI). *Ecological Economics*, 58(1), 119–133. doi:10.1016/j. ecolecon.2005.02.008

Wakeman, S. E., Larochelle, M. R., Ameli, O., Chaisson, C. E., McPheeters, J. T., Crown, W. H., . . . Sanghavi, D. M. (2020). Comparative effectiveness of different treatment pathways for opioid use disorder. *JAMA Network Open*, 3(2), e1920622–e1920622. doi:10.1001/jamanetworkopen.2019.20622

Wang, J., & Wang, X. (2012). *Structural equation modeling*. West Sussex: Wiley.

Wang, S., Chen, C.-C., Dai, C.-L., & Richardson, G. B. (2018). A call for, and beginner's guide to, measurement invariance testing in evolutionary psychology. *Evolutionary Psychological Science, 4*(2), 166–178. doi:10.1007/s40806-017-0125-5

Watson, J. (1979). *Nursing: The philosophy and science of caring* (1st ed.). Boston, MA: Little, Brown.

Watson, J. (1985). *Nursing: Human science and human care.* New York, NY: Appleton-Century-Crofts.

Watson, J. (2008a). *Nursing: The philosophy and science of caring.* Denver, CO: University Press of Colorado.

Watson, J. (Ed.) (2008b). *Assessing and measuring caring in nursing and health science* (2nd ed.). New York, NY: Springer.

Weiss, R. D., Griffin, M. L., Marcovitz, D. E., Hilton, B. T., Fitzmaurice, G. M., McHugh, R. K., & Carroll, K. M. (2019). Correlates of opioid abstinence in a 42-month posttreatment naturalistic follow-up study of prescription opioid dependence. *Journal of Clinical Psychiatry, 80*(2). doi:10.4088/JCP.18m12292

Wessel, S., & Manthey, M. (2015). Primary nursing: Person-centered care delivery system design. Minneapolis, MN: Creative Health Care Management.

West, B. (2007). *Where medicine went wrong: Rediscovering the path to complexity.* Singapore: World Scientific.

Wong, C. A., Recktenwald, A. J., Jones, M. L., Waterman, B. M., Bollini, M. L., & Dunagan, W. C. (2011). The cost of serious fall-related injuries at three Midwestern hospitals. *The Joint Commission Journal on Quality and Patient Safety, 37*(2), 81–87. doi:10.1016/s1553-7250(11)37010-9

Wu, F. M., Newman, J. M., Lasher, A., & Brody, A. A. (2013). Effects of initiating palliative care consultation in the emergency department on inpatient length of stay. *Journal of Palliative Medicine, 16*(11), 1362–1367. doi:10.1089/jpm.2012.03

Yancy, C. W., Jessup, M., Bozkurt, B., Butler, J., Casey, D. E., Colvin, M. M., . . . Westlake, C. (2017). 2017 ACC/AHA/HFSA focused update of the 2013 ACCF/AHA guideline for the management of heart failure: A report of the American College of Cardiology/American Heart Association task force on clinical practice guidelines and the Heart Failure Society of America. *Circulation,* 136(6), e137–e161. doi:10.1161/CIR.0000000000000509

Yancy, C. W., Jessup, M., Bozkurt, B., Butler, J., Casey, D. E., Drazner, M. H., . . . Wilkoff, B. L. (2013). 2013 ACCF/AHA guideline for the management of heart failure. *Circulation,* 128(16), e240–e327. doi:10.1161/CIR.0b13e31829e8776

Yip, W. K., Mordiffi, S. Z., Wong, H. C., & Ang, E. N. K. (2016). Development and validation of a simplified falls assessment tool in an acute care setting. *Journal of Nursing Care Quality,* 31(4), 310–317. doi:10.1097/ncq.0000000000000183

Zamirinejad, S., Hojjat, S. K., Moslem, A., Moghaddam Hosseini, V., & Akaberi, A. (2018). Predicting the risk of opioid use disorder based on early maladaptive schemas. *American Journal of Men's Health*, 12(2), 202–209. doi:10.1177/1557988317742230

Zimlichman, E., Henderson, D., Tamir, O., Franz, C., Song, P., Yamin, C. K., . . . Bates, D. W. (2013). Health care–associated infections: a meta-analysis of costs and financial impact on the US health care system. *JAMA Internal Medicine*, 173(22), 2039–2046. doi:10.1001/jamainternmed.2013.9763

Index

Using Predictive Analytics to Improve Healthcare Outcomes, First Edition.
Edited by John W. Nelson, Jayne Felgen, and Mary Ann Hozak.
© 2021 John Wiley & Sons, Inc. Published 2021 by John Wiley & Sons, Inc.